P9-CEC-911

THE SPANISH ARMADA

The Spanish Armada

The Experience of War
in 1588

FELIPE FERNÁNDEZ-ARMESTO

OXFORD UNIVERSITY PRESS
1988

Oxford University Press, Walton Street, Oxford OX2 6DP

Oxford New York Toronto
Delhi Bombay Calcutta Madras Karachi
Petaling Jaya Singapore Hong Kong Tokyo
Nairobi Dar es Salaam Cape Town
Melbourne Auckland

and associated companies in
Berlin Ibadan

Oxford is a trade mark of Oxford University Press

Published in the United States
by Oxford University Press, New York

British Library Cataloguing in Publication Data
Fernández-Armesto, Felipe
The Spanish Armada: the experience of
war in 1588.
1. Armada, 1588
I. Title
942.05′5   DA360
ISBN 0–19–822926–7

Library of Congress Cataloging in Publication Data
Fernández-Armesto, Felipe.
The Spanish Armada: the experience of war in 1588/Felipe
Fernández-Armesto.
Bibliography: p.
Includes index.
1. Armada, 1588.   2. Spain. Armada—History—16th century.
3. Spain—History—Philip II, 1556–1598.   4. Great Britain—History,
Naval—Tudors, 1485–1603.   I. Title.
DA360.F43 1988   946′.04—dc19   87–22914
ISBN 0–19–822926–7

Printed and bound by
Butler & Tanner Ltd,
Frome and London

Preface

SIR FRANCIS DRAKE, interrogating a Spanish prisoner just before the launching of the Armada, was told, with wild exaggeration, that the fleet contained over two hundred war ships. 'No es mucho,'—'That's not a lot,'—he is said to have remarked in reply.[1] At first glance, this looks like a typical confrontation between English understatement and Spanish pride. In the literary and historical tradition inspired by the Armada, however, these presumed national characteristics appear to have been mutually exchanged. On the Spanish side, apart from a flurry of activity by naval historians in the last century, and a good deal of work in progress now, reticence—dignified or embarrassed—has generally prevailed. If Spanish historians have succumbed at all to the supposed national vice of hyperbole, it has tended to be in exaggerating the extent of their own countrymen's defeat.[2] The English tradition, on the other hand, has been typified by some fairly loathsome chauvinism and by a pride in achievement unjustified—as I propose to argue—by the facts. I have painful memories of 'doing' the Armada as a boy in an English prep school in the 1950s, and of reading, with envy at the unquestionably English credentials of my contemporaries, some of the patriotic poetry it spawned. Austin Dobson's verses were a particular source of embarrassment to me—

> King Philip had vaunted his claims.
> He had sworn for a year he would sack us—

because to a boy with a name like mine the next lines had a peculiarly fateful resonance:

> With an army of heathenish names
> He was coming to faggot and stack us.
> Like the thieves of the sea he would track us
> And scatter us on the main,
> But we had bold Neptune to back us,
> And where are the galleons of Spain?

Yet this was one of the mildest, as well as one of the wittiest, works in a tradition warped by three distorting influences: the 'Black Legend' of

Spanish cruelty and fanaticism, which was based on misapprehensions about Spanish conduct in the New World in the sixteenth century and provided a specious explanation of the motives underlying the attempted invasion of England; the 'Whig interpretation' of English history, which represented the struggle against Spain as a quasi-constitutional conflict of liberty with despotism; and Protestant apologetics which saw the same war as a war of religion and in extreme cases—even in the hands of serious and accomplished scholars—represented the outcome of the Armada as a demonstration of the superiority of the Protestant religion. All these influences came together in the literature generated by the last Armada centenary celebrations of 1888 and particularly in the publications of the Protestant Alliance. Wylie's *History of Protestantism* was an especially virulent example in prose; but the most representative and most risible accounts were offered in verse or drama. Augustus Harris, who toured provincial centres with a so-called 'Historical Lecture', spreading the image of the Armada as a vector of allied popery and tyranny, was also the author of a *Romance of 1588*, a stage spectacular which culminates in a hair's breadth rescue for the English heroes from the clutches of an *auto de fe*. W. H. Smith's *The Spanish Armada* is perhaps most comprehensive in combining Protestantism, Whiggery, and moral outrage, recoiling from the stuff of the Black Legend with evident relish and affected distaste. 'How brilliant with delusive glow,' the author exclaims, 'Glamour above and death below / Spain's glories past have been!' The events of 1588 are seen as a victory for English freedom over Spanish absolutism, which inflicted 'red, gaping wounds' on 'West Indians, Piedmontese and Moors'. The writers of 1888 were drawing on a long tradition. Harris quoted freely from Protestant pasquinades of the seventeenth century. The tedium of Macaulay's poem in praise of the beacons that flashed warnings of the Armada across the country is something which in a previous generation 'every schoolboy' knew.[3]

Apart, perhaps, from the Black Legend itself, the Armada has played a bigger part than any other single influence in shaping English perceptions of Spain. It has also, I think, had a disproportionate part in defining English self-awareness. Two elements of the English self-image—sang-froid, and a preference for the underdog's role—are linked in almost every English mind with myths of the Armada: the supposedly plucky little ships worsting the Spanish giants, like David against Goliath; and the probably apocryphal game of bowls Drake played on Plymouth Hoe. The cult of the reign of Elizabeth as England's age of national greatness—quite at variance with the facts, since England achieved more, by almost any measure, in relation to her rivals in, say, the fourteenth century or the

nineteenth—is intimately connected with the image of Elizabeth address-
ing her troops at Tilbury in words so well known in England that they
can almost be described as a recollection of folk-memory. That image,
and its importance in English life, is brilliantly expressed and brilliantly
satirized in E. F. Benson's portrait of a village pageant, in which his
socially ambitious heroine, Lucia, establishes her local ascendancy by
appearing in that role and declaiming that speech. Neither in England nor
Spain is there any longer much interest in perpetuating adversarial myths
or sustaining mutually pejorative images. The main outstanding conflicts
of interest between the two peoples—over EEC quotas and the obsolescent
anomaly of Gibraltar—are, it is generally agreed, best resolved by means
of enhanced mutual understanding. The Armada is therefore bound to
be seen differently in England now, in its four hundredth anniversary
year, from the way it was presented at the time of the last celebrations. The episode seems most conspicuous today, not as a conflict of
religions or value systems, but as a passage of arms in a shared past,
of two peoples jointly distinguished from their other partners in the
European alliance by a common maritime imperial experience, which
has often brought them into conflict, but has left them with a common
heritage.

My purpose in writing this book, however, has not been to tilt at
creaking windmills of bygone myth, which, the reader may feel, have
anyway little wind left in their sails. I want rather to expose an aspect of
the history of the Armada which, perhaps because of the strength of the
traditions to which I have referred, has been neglected even in scholarly
accounts; and to explore some problems which have often been raised but
never resolved. I see the Armada not as the extraordinary and crucially
important event so often depicted, but—in most respects—as a typical
episode of sixteenth-century warfare, of no great importance in itself, but
of interest as a case-study of the disasters of war. The common experience
of those who took part has often been pointed out in one respect: the
common suffering, from sickness and want, of the men on both sides in
the aftermath of the Armada. In the present essay, I have tried to identify
elements of common experience at every stage of the story: the logistical
efforts in Spain and England; the chaotic strategic preparations and
muddled thinking at high levels of command; the frustration of the
ineffective tactics employed on both sides; the gruelling and disappointing
experience both forces had of battle; the common war waged by both
fleets against the weather, until the English desisted and the Armada was
left to struggle on alone. I describe common experiences and argue for an
evenly balanced outcome; in particular, I challenge the notion of a Spanish

defeat at English hands, or, at least, argue that it can be accepted only in a heavily qualified sense.

The specific problems of the history of the Armada to which I try to make a contribution in these pages are: first, what was the purpose of the Armada and how, in Spanish strategic thinking, was it intended that this purpose should be achieved? All historians of the Armada have identified elements of incoherence, irrationality, and internal contradiction in the Spanish plans, and rightly so; but previous accounts, it seems to me, have made Spanish plans seem so foolish as to be incredible. I have tried to reconstruct the evolution of Spanish planning in intelligible stages and to show that in the end a theoretically workable plan was extemporized, though the means were lacking to bring it off. I also take the view—now unlikely to be widely disputed—that the Armada was seriously intended to undertake an invasion, if not necessarily to complete the conquest, of England.

Secondly, I join the traditional debate about the technical and tactical differences between the two sides. I find that these were substantial, but argue that none of them made much difference to the outcome of the encounter. I take the opportunity to question the view that the Armada fights marked any major new departure in the history of naval tactics or strategy. I review all the explanations I can think of or have read about for the ineffectiveness of the tactics and ordnance of both sides and make what I believe are new suggestions. Thirdly, or more generally, I raise the problem of why the Armada failed. To some extent, I try to represent this as a *question mal posée*, attributing to the Armada rather more success than is commonly acknowledged; in common with what seems now to be the prevailing trend among scholars, I emphasize the role of the weather, and minimize that of the English and Dutch and of the supposed inherent weaknesses of the Armada.

Despite my insistence on seeing the Armada episode as a common experience, the Spaniards take up more of this book—as they do of all books on the Armada—than the English. This is because there are more Spanish than English sources, and because the war against the weather, to which I assign considerable attention and space, was waged longer by the Armada than by the English fleet. I have left out the diplomatic prelude to the Armada campaign: this was brilliantly—and almost com-prehensively—covered by Professor Garrett Mattingly's *Defeat of the Spanish Armada* (London, 1959) and is being done again, with the resources of recent research, by Dr Mía Rodríguez-Salgado for her intro-duction to the catalogue of the forthcoming Armada exhibition at the National Maritime and Ulster Museums. Almost all books on the subject have tried to provide a narrative of events, none entirely satisfactory; but I have nothing new to contribute in that respect and therefore make no such attempt here: events from 30 July to 5 August 1588, are dealt with analytically in one chapter; those of 6 to 8 August in another. In the last two chapters, though I am concerned to establish the broad chronology of losses of ships, I adopt a narrative approach only in covering the events of the rest of August. In the last chapter, the categories into which I group the losses of September 1588 are based broadly on those established by N. Fallon, *The Armada in Ireland* (London, 1978).

There has never been a better or worse time to write about the Armada: never better, because the approach of the fourth centenary has aroused interest and stimulated scholarly research. The present essay is unashamedly a *pièce d'occasion*, produced by Oxford University Press as a contribution to the commemoration. Never worse, because all the new work is sure to yield insights which will be gradually revealed and slowly absorbed; they will modify some judgements and correct some errors which are bound to be made now. In particular, three projects at present in hand are likely, taken together, to have a transforming effect on the way the Armada will be seen in future. A team of Spanish scholars is at work on the great mass of unexplored and ill-explored Spanish sources and will soon begin to publish its findings in a series of monographs. The National Maritime Museum, Greenwich, and the Ulster Museum are together preparing a commemorative exhibition for the fourth centenary of the Armada, which will bring together the biggest collection of sources of all kinds ever assembled, including most of those yielded by the work of marine archaeologists. And Professor Geoffrey Parker and Dr Colin Martin have united a great deal of previously neglected material, from Spanish and Dutch archives and from the sea-bed, for a television series and accompanying book. In the longer term, it must be hoped that the Medina Sidonia archives, which disappeared from view a few years ago, after being only very briefly open to scholarly inspection, will resurface and yield further information. In the mean time, however, there remains a great mass of available material which is well worked but by no means exhausted, and from which many new perceptions can be wrested and many new views exemplified.[4]

Circumstances have dictated that this book be written in the depths of rural Lincolnshire, far from libraries, archives, and the conversation of other scholars. I mention this not to excuse its shortcomings but to explain its form, which is that of a discursive essay rather than an exhaustive compendium or research monograph. This means that I have had to rely to some extent on intermediate sources—Hume's abstracts for many

Spanish documents, for instance, and Van der Essen's for Parma's correspondence. Selective checks have convinced me that the results are reliable. I have, at least, had the inspiration of an Armada panel-painting in a parish church and the Bertie tombs at Spilsby close at hand.

F.F.-A.

Partney House, Lincolnshire–St Antony's College, Oxford
January–June 1987

Contents

Maps

One

The Sinews of War

A SKINNY poet on a bony mule was an unremarkable sight in late sixteenth-century Spain. On this September day, however, moving slowly under the hot sun eastwards from Seville into the rich grainlands of Lower Andalusia, the rider and his beast presented curious features and may have drawn apprehensive glances. Poets—as Cervantes said—are normally baggageless. But this one's saddlebags bulged with bulky documents in unpoetical language, authorizing him, on the king's behalf, to search men's houses and impound their goods. For the year was 1587. The poet was Cervantes himself. And his commission was to raise purveyance for the Spanish monarchy's most ambitious undertaking since the conquest of Peru.[1]

The Spanish Armada called forth enormous and in some ways unprecedented logistical efforts on both sides. The case history of Cervantes vividly illustrates the typical muddle and waste, sacrifice and pain, strained loyalties and social conflicts exacted or imposed by the slowly grinding process of supply. The duties of a purveyancer were irksome to the poet and unworthy of his character but by no means foreign to his experience. Though poetry was a fashionable taste, and even a socially desirable occupation in the Andalusia of his day, aristocratic patrons were neither numerous nor generous enough to sustain literature unaided. An office of profit or pay under the crown or in the church was the normal recourse for a man of letters. For men of Cervantes's class, well educated and underemployed, the additional official activity associated with the launching of a great fleet was a bonanza. His own temperament was unsuited to business and Cervantes made relatively little of his opportunity; but at least it lifted him, and many others like him, out of penury for a while. He had made a special trip from Toledo to Seville to secure a commission and spent the four months of that summer—an exceptionally hot one, even by the standards of southern Spain—sweating amid the queues of suppliants at the purveyor-general's door, begging for a place.

Expectations of gross inefficiency were built into the system of purveyance of which Cervantes formed a part. It was assumed that the provisions, however much was garnered, would never be sufficient for

The most useful visual record of the Armada was made by Robert Adams and engraved on copper by Augustine Ryther, who also made marine instruments. As the title-page illustrates, it is a work of remarkable beauty and artistic originality as well as an important historical work. The bold *trompe-l'œil* illusionism of the strapwork design is worthy of a Max Escher. Notice the dangling anchors.

the job, and that in any case a vast amount would be lost by natural spoiling and peculation. The purveyor-general's agents, of whom Cervantes was one, were given quotas of grain, fodder, and oil to be raised in each place but were left, in effect, to gather as much as they could, restrained chiefly by the system's own in-built inertia, obstruction, delays, and denials. They paid for the goods they solicited or seized, as they were paid themselves, with promises from the crown. The amount of actual cash that changed hands was barely enough to maintain confidence in the system, which depended, ultimately, on a remarkably elastic capacity for patriotic sacrifice in an already overtaxed community. The inhabitants of the first town Cervantes visited had not yet been paid for the provisions levied from their previous harvest; during Cervantes's tenure, all the money he received went on paying the expenses of his operation, sparing nothing with which to reimburse the producers. Of his own salary of twelve *reales* a day he received nothing until his commission had ceased, and even then—according to standard practice—his accounts were revised by the bureaucracy in the crown's favour and the bulk even of the revised account was never paid. For the eighteen months his job lasted, however, he lived comfortably at the expense of the king and his fellow citizens. By the end of that time, of course, the Armada was in tatters and most of the supplies he had gathered had rotted away without ever leaving store.

On paper, he went armed with sweeping powers. He left Seville with a rod of office, received from the hands of the purveyor-general's deputy, and credentials which detailed his intimidating powers: locks could be forced, private premises violated, goods distrained, transport halted and seized, individuals fined or imprisoned, and every requisite impounded, if necessary, down to baskets and brooms, paper and ink. In reality, the Spanish state was too inefficient, too riven by competing jurisdictions, and too little practised in absolutism for the system to work without consent. Cervantes's well-documented efforts to raise grain and oil for the Armada in Écija and La Rambla reveal every stage of the slow, wary process by which supplies were coaxed from local communities, jealous of their privileges and skilful in litigation, but ultimately willing to render reasonable service to the crown. First the inhabitants of Écija pleaded an

The young Sir Edward Hoby (see p. 206), painted five years before the Armada at a time of increasing tension with Spain. The approach of war is alluded to in the allegory in the top right-hand corner. The queen emerges from a stronghold to inspect arms which are gleaming but lightly covered with a thin net. The legend reads 'They are laid aside but not rusted.'

inability to receive Cervantes because of a *fiesta*: he might return, they suggested, in three weeks. Then the town council, reluctantly obliged to meet him, appealed to the terms of the crown's own instructions to purveyors, exempting seed-corn and local food; they claimed the harvest had been savaged by bad weather and that the people were already suffering severe want because of sums owed by the crown from the last purveyance; they offered to compound for their quota and then withdrew the offer; they made repeated appeals to Seville and Madrid on the grounds that Cervantes was exceeding his powers or even stealing the wheat. They procured an opinion from a local Dominican that Cervantes's proceedings were contrary to divine law. They took advantage of the fact that he had impounded grain belonging to the church by invoking first ecclesiastical censure, then full excommunication—twice incurred by Cervantes in the course of his tour of duty. They demanded and were given guarantors—locally and at court—for the money owed them.

To these traditional methods of tempering the exactions of the crown, Cervantes replied with the equally traditional countermeasures of the bureaucracy. He joined in the ritual negotiations to compound for the quota; he summoned his superior, the deputy purveyor-general, for an intimidating visitation of the town; he paraded around the town with rod in hand and sergeants in attendance; he made token seizures of property, forcing locks and sealing granaries. Such were the 'excesses in requisitioning in Andalusia, by the justices of those places, the captains, officials and other ministers' of which the people complained to the king and which the king reported to the commander-in-chief. Ultimately, as usual, it was all settled by accommodation. The inhabitants could command such formidable methods of delay that all the supplies would be spoiled and useless by the time recourse to the law had been exhausted. In the end, in late June 1588, all was amicably settled. Local notables and churchmen were paid for what had been requisitioned from the previous harvest. The town pronounced itself satisfied with Cervantes's guarantees of payment for the newly requisitioned produce, and milling might even have begun in time to ship some biscuit to the fleet in Corunna, had new demands from Seville for further supplies not caused the cycle of disputes to begin all over again. Cervantes spent the heat of the summer in the barns and on the threshing-floors, minutely inspecting the preparatory work in the stifling, choking atmosphere. No detail that a pettifogging bureaucracy could envisage was overlooked in his voluminous written instructions. He was to identify the best mills; store the wheat where it would not be damaged; divide it into equal lots, test samples at random, having them cleaned and winnowed in the presence of the legal representatives of the town; the wheat, the chaff, the dust from the threshing-floor were to be separately weighed and inventoried, 'noting everything in detail'; and the notarized records were to be sent on to Seville, where an abstract would be made for Madrid. By mid-July, nearly four-fifths of the grain Cervantes had collected was unusable. Milling at last began in earnest in August, when the Armada was in the throes of battle, beyond the reach of further supplies.

Meanwhile, what was not spoiling in Spanish granaries was rotting in the ships' holds. The storm that drove the departing Armada into Corunna in June provided an opportunity for the inspection of supplies. Many had been garnered and delivered so long before that they were already bad. One of the most senior commanders, Don Pedro de Valdés, found his own squadron had its full three months' rations of biscuit but that it was already beginning to decay. There was ample wine, but the bacon, cheese, salt fish, sardines, and such vegetables as he had were all rotting 'and of very little use'. There was a dearth of oil. Fresh meat would set the men up while in port at relatively modest cost and conserve the remaining stores—but in a sense, of course, this would only compound the problem by delaying the consumption of decaying goods. All the captains agreed that, except for the rice, pulses, wines, and some of the biscuit, 'the victuals were of no use whatever, for the men would not eat them'. According to the inspector-general, Don Jorge Manrique, the fish and bacon was 'all spoilt and rotten, as it has been on board for so long'. Except in Valdés's squadron, the fleet was well short of its intended three months' rations—there was barely enough for two months, according to the commander-in-chief.

Thus, even before the Armada had left for England, waste was eating its way into the stores. The distances over which supplies had to be gathered, the delays to which departure was subjected, and the slow rituals of purveyance in a pre-absolutist state obliged victuals to be hoarded while they spoiled. In Corunna, for instance, as one of the most senior squadron commanders, Juan Martínez de Recalde, complained, while the fleet gradually reassembled, 'days were spent discussing what ought to be done, making the manger even smaller than it is'. The wastage spread to the men. On 24th June, the Duke of Medina Sidonia, overall commander of the Armada, warned the king, 'The crews are sick and dying fast, on account of the bad provisions and the danger from them, as I have constantly reported to Your Majesty'. The duke was being unnecessarily alarmist—the violent storm, the dispersal of the fleet, and the delay in Corunna had all evidently frayed his nerves and dispelled his customary unflappability; on 6 July he admitted that the crisis was over: the fever

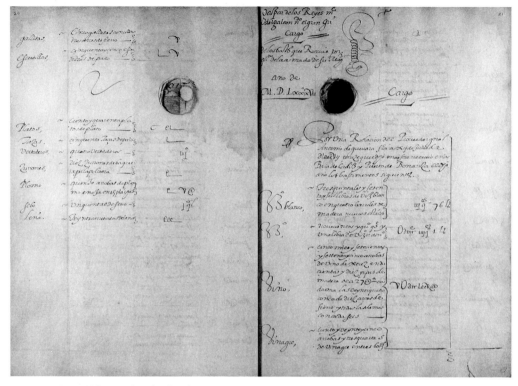

Leaves from a 'Book of Ladings, Provisions, and Stores made for the Captains, Masters, and Owners of the Ships and other Vessels that are to serve in the Armada of His Majesty, this year of 1587' by the purveyor Bernabé de Pedroso. The right-hand page lists hard tack, wine, and vinegar aboard the *Gran Grín*, here inaccurately described as a galleon; small-arms munitions are listed on the left. Notice the hole piercing the manuscript, probably for convenience in hanging the book from a hook aboard ship.

cases were recovering in shore hospital and he was expecting to sail with something approaching full strength. But his anxiety shows how, even before he began his battle with the English, his morale was already reeling under the impact of his struggle with the fleet's intractable logistic problems. He knew them all too well. As captain-general of Andalusia he had borne, in effect, the main responsibility for organizing the transmission of men, ships, and supplies to the Armada, long before assuming supreme command.[2]

Yet, in view of the scale of the venture, the Spanish operation is more surprising for its efficiency than its failures; Medina Sidonia's promotion, indeed, was official recognition of this. Drake did his adversaries no more than justice when he reported, well over a year before the Armada sailed, that 'there was never heard of so great a preparation as the King of Spain hath and doth continually prepare for an invasion'. The Spanish monarchy had prepared bigger undertakings on paper in the past. A

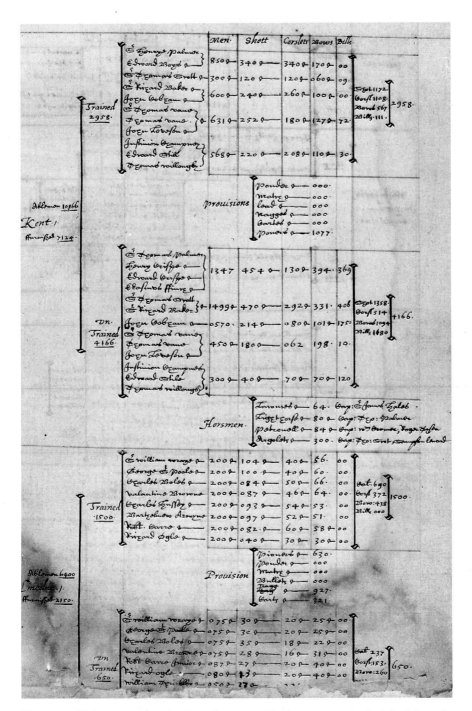

The cost of defence against the Armada was meticulously computed. A leaf from Sir John Hawkins's 'Estimate of the Charge of Her Majesties Army prepared against the Spaniard', showing the pay and cost of stores (misleadingly called 'provisions') for the trained bands and militia of Kent and Lincolnshire.

planned invasion of Barbary of about 1538 envisaged a fleet of 384 ships, totalling 130,000 tons, manned by 15,320 sailors and 60,000 oarsmen, to convey an army of 84,700 men with 8,000 horses. Only in Spain could such a scheme be hatched. The Armada was at first conceived on an even more grandiose scale. When Philip II asked the leading advocate of the 'enterprise of England', the Marquess of Santa Cruz, to estimate what would be required for the campaign, the old admiral called for a fleet of more than 550 sail, including 156 great ships and forty or fifty galleys with 30,000 sailors and 64,000 soldiers. When the Armada was formed, its strength and resources would amount to little more than a third of what Santa Cruz had originally envisaged.[3]

The phantom Armada of Santa Cruz was so impossibly vast that many historians have supposed that it cannot seriously have been intended. The marquess, it is argued, must have lost confidence in the enterprise of England and set his demands unacceptably high in order to get the project shelved without loss of face. Certainly, as time went on, his enthusiasm seems, from other evidence, to have waned. Yet he made out a rational case for saying that his estimates, though unprecedented, were not beyond the monarchy's capacity, and he coined what might have been the motto of the undertaking, 'In my view, this business has to be taken in hand with a gallant resolve, and leave the rest to God.' After all, his master, Philip II, never hesitated to invoke divine aid. Meanwhile, two changes of policy or strategy occurred to rescue the plan, or, rather, to reduce it to manageable proportions. The first, and much the lesser, was loss of faith in galleys as suitable vessels in northern waters. As Santa Cruz had called for a force of up to fifty galleys—almost a tenth of the whole—this alone represented a significant deduction. The second innovation was the decision to ferry only a small part of the invasion force from Spain, relying instead on the use of some of the Army of Flanders. This meant that the fleet from Spain could be limited to the dimensions of a fighting task force, with a convoy of reinforcements and supplies. Ironically, in reducing the scale of the enterprise to manageable proportions, these two changes foredoomed it to fail. Galleys—as we shall see—might have made a vital tactical contribution, had they been available. And, because it lacked adequate transports and escorts, the Army of Flanders was to prove useless when it was needed.[4]

Even the truncated Armada was, in practical terms, the biggest expedition ever mounted from Spain—vast not just in bulk and tonnage and manpower, but also in the distance over which it had to operate and the time it would have to remain afloat. The fleet, as it actually took shape, comprised 130 ships, by most accounts, including thirty-five which

The Marquess of Santa Cruz embodies the virtues of gravity and propriety esteemed at the court of the mature Philip II. Note the restrained ornament of his armour and the cross of the Order of Santiago in an intaglio on his breast. The inscription, summarizing the achievements of his life, places two successful actions against English shipping in defence of the Indies treasure fleets at the head of the list.

might be classed as first-class fighting ships, twenty of them galleons, which were the vessels most favoured for ocean warfare. Sixty-eight other great ships were large enough to serve as armed auxiliaries, supply vessels, or, in two cases, hospital ships. By the time they sailed from Lisbon, they bore—on paper, at least—all told, 110,000 quintals of biscuit, 11,117 pipes of wines, 11,000 pipes of water, 6,000 quintals each of bacon and salt fish, 3,000 quintals of hard cheese, 4,000 quintals of rice, and at least 6,000 bushels of beans. To have united such a formidable force was a triumph of organization. The provisions made for ordnance, powder, shot, and stores were little short of miraculous, though it must be remembered that they were not all meant to be expended at sea but also to serve subsequently, if all went well, in a major land campaign. The fleet carried 123,790 rounds of great shot and well over half a million rounds of best-quality powder, for guns and small arms. There were nearly 2,500 pieces of great ordnance, counting the great siege guns stowed in the holds, 7,000 arquebuses, 6,170 grenades and other incendiary devices, with 11,128 pikes. The nominal strength, at full muster, would have exceeded 30,000 men, over 18,000 of them regular soldiers.[5]

All this was merely what was laid on at one of two great termini of the operation. In the Netherlands another army had to be gathered, of 16,000 men, and another flotilla built virtually from scratch, of barges to ferry the men to England and flyboats or pinnaces to escort them out to deep water. That the flotilla proved inadequate in the event is less surprising than that it should have been called into being, in such a relatively short time, at all. It was not until September 1587 that its commander, the Prince of Parma, possessed a fully secure harbour. The scale of the logistic effort required may be gauged from a single vivid illustration: in March 1588, Parma's ovens were baking 50,000 loaves a day. The intractability of the problems around Parma's camp may be judged by the results. The boat-building programme on which he embarked to provide transports for his fleet foundered on shortage of suitable timber and willing labour. The local seamen and workmen, in the words of a Spanish soldier, 'undisguisedly directed their energies not to serve His Majesty, for that is not their aim, but to waste his substance and lengthen the duration of the war'. The workforce could not be reduced to military discipline, nor compelled to work without pay. Green and rotten timbers found their way into hulls that therefore proved unserviceable—perhaps, as the Spaniards suspected—through sabotage. Most of the barges Parma eventually got together were hired or commandeered from canal- and river-fleets. Most of his fighting boats were borrowed from neutral ports. Even these expedients were inadequate in the event. When representatives of the Armada arrived in Parma's camp to co-ordinate operations, they claimed to be bitterly disappointed by the laggard state of readiness they found. Yet this was only because of their excessive expectations. Parma insisted at the time that he could be ready in six days, and even Rodrigo Tellez, who was reporting on the Armada's behalf, believed a fortnight might suffice. It was not good enough, but, in the circumstances, it was remarkably good. To maintain a large army for so long, in so near a state of readiness, was an achievement equalled nowhere in Europe at the time outside the Spanish monarchy. Like other aspects of the preparation, if not the application, of the Spanish Armada campaign, it was a Spanish triumph of organization over adverse circumstances.[6]

How was the Spanish logistic triumph possible? Only a global monarchy could command the necessary resources; only a bureaucratic state could mobilize them; only painful expedients could pay for them. When Portugal had raised an army of 30,000 men in 1579 the country had been beggared as a result. King Sebastian had borrowed the equivalent of 400,000 ducats from German bankers at eight per cent; he had extorted 240,000 more from his 'New Christian' subjects, descendants of converts from Judaism, in exchange for a ten-year stay on Inquisitorial confiscations of property; he had sold crusading bulls granted for the purpose by the pope; he had levied aids from nobles, churchmen, traders, and towns; he sold 130,000 ducats' worth of silver which was awaiting coining in the royal mints and he raided the coffers of endowed charities. Yet these sums were insignificant in relation to those demanded by the Armada. Santa Cruz had costed his original plan at 3,800,000 ducats—a modest sum, he argued, in relation to the total turnover of the monarchy. But he reckoned without the cumulative cost over the many months it took to build up the fleet and organize the logistic support and, ultimately, to co-ordinate intractable operations in distant theatres with imponderables such as weather and wind and the movements of the enemy. In the course of 1587, it was thought at court that the Armada had cost ten million ducats. It cost 700,000 more for every month spent in port. Even the world's mightiest monarchy could not sustain this sort of expenditure without taking a severe toll of its fiscal resources.[7]

Luckily for Philip II, the yield of silver from the Indies was relatively high in the years from 1585 to 1591: Medina Sidonia gave good advice to the king when he urged repeatedly that the safety of the bullion fleets should take precedence even over the Armada, for on the one depended the other. The rest of the money could only be raised by sacrificing some long-standing principles of royal policy. Jurisdiction was sold off to the nobility with almost reckless abandon; the rights the crown had once struggled to gain in the lands of the knightly orders were alienated to the tune of three and a half million ducats in the second half of the 1580s. Philip temporarily gave up his efforts to cut back on the sale of offices, and with them he determinedly sold titles of nobility, common lands, trading privileges, and rights of lordship. A 'seigneurial reaction' was, to some extent, under way in any case; and the Armada was only one episode in a long history of costly wars that caused the Spanish Habsburgs to forfeit much of their domestic power. But the late 1580s do seem to have been a decisive catalyst. The Armada cost the Spanish monarchy, perhaps, its last chance of establishing absolutism. And still the cash was not enough. At the end of February 1588, Parma sent Philip one of his dry appeals—he left the excited pleading to Medina Sidonia:

> I will only remind you now of the importance of the question of money, and of its timely supply, both for the purpose of fulfilling the engagements entered into with merchants and for the maintenance of the preparations already made, if we are to avoid the trouble and inconvenience which will otherwise, as usual, be caused.

'Trouble and inconvenience' was a restrained euphemism; campaigns collapsed, armies mutinied, and strategies reversed were the common consequences of the monarchy's chronic belatedness over pay. In the margin of this reminder, the king noted characteristically, 'This is the matter which gives me most anxiety. But in this as in all things, unless the weather frustrates us, I have firm hope in God.'[8]

Cash was useless without supplies to spend it on. Nothing so vividly illustrates the cosmopolitan nature of the Spanish monarchy of the time than the diversity of the Armada's provender. Salt meat and lard were traditionally provided by Galicia, Ronda, and Málaga, sometimes by Sardinia and Naples. Salt anchovies and sardines were supplied by Galicia and Andalusia, oil by Andalusia, cheese by Sardinia, and rice by Valencia and Lombardy. Sicilian pulses, chick-peas, and beans, were vital sources of protein—scarcely more expensive than wheaten hard tack, to which all the grainlands of the monarchy on the near side of the Atlantic contributed. According to an English intelligence report of March 1587, 56,000 bushels of biscuit had been drawn from Burgos and Old Castile, 40,000 from Naples, 25,000 from Lower Andalusia, 27,557 from Upper Andalusia, 12,000 from Málaga, 12,000 from Murcia, and 12,000 from the Canary Islands. Wines, because they kept well, could be drawn from far afield with ease. The Andalusian flagship *Nuestra Señora del Rosario* had wines of Jeréz, Crete, and Ribadavia on board. Medina Sidonia's instructions on rations mention Condado and Lisbon wines, the former from his own family's oldest vineyards, in western Andalusia, to be consumed first because they travelled poorly, followed by other Andalusian wines and then the wines of Jeréz and Crete which 'bear a sea voyage better'. Cretan wine, being the strongest, had to be served in a

Lisbon, 'noblest entrepot of many islands and of Africa and of America', with a list of 140 public monuments. The shipping in the harbour includes a galleon, a hulk, and caravels.

slightly smaller measure than the rest, mixed with a double quantity of water. This must have been good malvasia, such as had been grown for export for centuries by the Venetians and Genoese.[9]

Ships were gathered in Lisbon by fair means and foul. Spain was a maritime Janus, with enormous Mediterranean responsibilities. Her access to shipping suitable for Atlantic and northern warfare was therefore limited. The kernel of the Armada was formed of galleons, ships of the fastest and most weatherly build available. Philip II had nineteen seaworthy galleons at his disposal—six fewer, and slightly smaller on average, than those of his adversary. Nine of them, including most of the biggest and most heavily gunned, belonged to the crown of Portugal and formed the guard of Portugal's royal ocean-going trade. There were ten Castilian galleons, purpose-built to guard Spain's bullion route: they have to be rated as fighting vessels of high quality, but in some respects they were inferior to their sister ships from Portugal. Newer and trimmer, they were also smaller: only their flagship, the *San Cristóbal*, with 36 guns, was in the same class as a bearer of ordnance with the Portuguese galleons; the other Castilian galleons had a relatively modest capacity of

24 guns each. They carried a comparatively high quantity of shot: 60 rounds per gun, against an average of 50 for the Portuguese ships, but, on the whole, the guns were of lighter weight of shot aboard the Castilian vessels. The Castilians' total charge of powder was much inferior, though it is dangerous to draw any firm conclusion from this, because an unknown proportion of the powder, vessel by vessel (for we know of only one case), may have been earmarked for small arms and the siege train. Certainly, the Castilian galleons played a more modest part in the battles fought by the Armada than their Portuguese counterparts. Of the ships of the Castilian squadron, the most widely reported for gallantry in surviving accounts was not a galleon at all, but the converted merchantman, *Nuestra Señora de Begoña*; it must be added, however, that she was a fairly trim vessel, sometimes mistaken for a galleon, and more heavily gunned than any other in the squadron after the flagship.

The galleons of Spain and Portugal were supplemented—on the most generous criteria—by a total of perhaps sixteen reputedly first-rate fighting ships, all but one of which were armed merchantmen. Five of them were of Mediterranean build and therefore of doubtful pedigree for the heavy seas the Armada was to traverse. Of these, only one, in the event, was to prove thoroughly capable, the almost brand-new galleon, universally admired, built by the Duke of Tuscany for Florence's part in the spice trade, and equipped with 52 brass guns. The *San Francesco* or, as she was known in Spain, the *Florencia*, was probably the most able and almost certainly the best-equipped ship in either fleet. She served the king of Spain, with grudging Florentine consent, in the Azores voyage of 1587, but the Florentines were reluctant to lend her for a purpose as hazardous as that of the Armada. The Marquess of Santa Cruz could not risk a diplomatic incident by impounding her, but her presence in Lisbon on a commercial voyage constituted an irresistible temptation. The Spaniards obtained the use of her simply by delaying her departure until her captain was constrained by the sheer hopelessness of his position to accept Spanish orders. Captain Niccolò Bartoli may have felt, in any case, that his ship was wasted on Mediterranean pirates, who constituted his other potential foe, and seems to have welcomed a chance of some real action against the English with a 'Catholic' fleet. It cost him his life. 'A few days after returning to port,' the Florentine ambassador reported on 15 October 1588, 'he died of a fever which had taken hold of him after the terrible journey and exertions he endured.' He need not, however, have felt any disappointment in the performance of his ship. Transferred from the squadron of her fellow Levanters to the flagship's own company, she was in the forefront of battle. Even before her return, Medina Sidonia's messenger assured the ambassador 'that the galleon ... has given the best account of herself and has turned out to be the finest vessel they had'.[10]

More subterfuge, of a different kind, was practised to get shipping from the people of Dubrovnik, then still known as Ragusa, which was famous for its shipyards. They produced large merchantmen, ideal both for supply ships and for adaptation as fighting carracks. Like Florence, Dubrovnik was a wary ally of the king of Spain. She was unable to make a contribution to the Armada openly, because of the constant need to keep Spain's Turkish foes appeased; at the same time, she was induced by commercial ties and Catholic fellow-feeling to look favourably on Spanish needs. The Ragusan contingent, therefore, joined the Armada, as it were, in disguise. Dubrovnik's ambassador to the Turks admitted that 'eight or nine' ships from his city had sailed with the Armada, but claimed that all had been 'requisitioned' by arbitrary Spanish action. Conniving in the ruse, the Spaniards gave the Ragusan ships new names, devised to conceal their provenance and to give the impression that they were of Italian origin. Thus the *Brod Martolosi*, apparently commandeered in Palermo in 1586, became the *San Juan de Sicilia*: she shared the Armada's greatest days of heroism and disaster. Obliged by battle damage to straggle on the journey home, she was driven off Islay on 23 September 1588 and thence to Tobermory Bay; Lachlan MacLean, the Lord of Mull, succoured her in exchange for help in his wars against his neighbours but planned treacherously to seize her; she would have escaped, but an English agent fired her magazine and she perished with the captain and most hands, only five of her native crew escaping. Other Ragusans included the *San Nicolás*, formerly the *Sveti Nikola*, whose captain, Marin Prodanelić, according to a letter to his brother from a correspondent in Lisbon, 'perished on the cliffs of Ireland'. There was also a ship known as the *Santa María*—not more precisely identifiable with any certainty among the many ships of that name but perhaps the *Santa María de Visón*—which, commandeered while carrying wheat from Sicily to Genoa, became the subject of a claim for damages sustained 'while she served in the English affair'. Of all the participants from Dubrovnik, the most committed were the Iveglia family—big shipbuilders and merchants with a long record of obligation towards Spain. Petar Iveglia was said to have supplied a dozen ships for the Spanish service. He had two nephews aboard the Armada—the name of one of them is recorded, in a garbled but unmistakable form, by his English captors. One of his ships, the *Presveta Anuncijata*, according to the deposition of a Spanish seaman who made it back to Laredo, was scuttled in the mouth of the Shannon, probably in Scattery Roads. Nothing illustrates better than the histories of these

Ragusan ships how far and wide the Armada was painfully gathered and pitifully scattered.[11]

From Venice and Genoa some large supply ships were stealthily borrowed or brazenly seized, to be adapted for battle. The *Trinidad Valencera* was commandeered from Venice in a Sicilian port in 1587 to transport troops to Lisbon, then impounded as a potential fighting ship. Her complement of guns was increased from 32 to 42 and her great bulk—for she was one of the four or five largest vessels in the Armada by capacity—was used to stow part of the siege train. In February 1588 the Venetian argosies *Regazona* and *Lavia* were requisitioned, in the sanguine expectation that 'the Venetian ships are so powerful that they can give battle to ten or twelve English ships'. Considered as troop-carriers, that may have been true of them, but, with the exception of the *Regazona*, they were to prove relatively frail and unweatherly ships, prominent in the fighting but vulnerable to its scars and susceptible to the rigours of northern seas. During the June storm that forced the Armada into northern Spanish ports, the *Trinidad Valencera*—at 42 guns, including some of the heaviest, one of the most powerful ships in the fleet, on which much reliance would be placed—exhibited a disturbing tendency to run before the wind. The Genoese carrack of 35 guns, the *Santa María*, known from her figurehead as the *Rata Encoronada*, later revealed the same defect. These two, together with some of the Ragusans, and closely followed by the *Lavia*, were all quickly to fall behind the rest of the fleet when the Armada reached stormy waters on its return voyage. There were other, more modest Levantine ships in the Armada, though it is hard to identify them individually. At least one pinnace, as we shall see (p. 244 below) was probably of Ragusan build, and there was a middle-sized Genoese vessel, of 400 tons by Spanish reckoning, with which a cabin boy called Gian'antonio di Manona came to Spain with his father, who was appointed pilot of the Biscayan *Santa María de la Rosa*: both would die tragically when the ship sank in the Blasket Sound, the father allegedly killed in error by a Spaniard who suspected him of treason, the son cynically executed by his English captors (below, pp. 243, 267).

Most of the balance of the Spanish fleet was probably drawn from within the king of Spain's own dominions, though some supply ships were chartered from abroad. We know, in particular, of four Hanseatic hulks: the provenance of the *Gran Grifón*, which came from Rostock, was betrayed by a note in Lord Burghley's hand in his own copy of the Armada's battle order. Those of the *Barca de Amburg* and *Barca de Anzique*—from Hamburg and Gdańsk respectively—are revealed by their names. The *Falcó Blanco Mayor* survived the voyage of the Armada only to be captured by the

English on her way home in 1589. Baltic or north German hulks carried supplies into Lisbon in course of the preparations for the Armada. But the squat merchantmen that bore the name of hulks were built within Spain, too, in Galician and Biscayan shipyards. Apart from hulks, which were often well armed but which were not normally expected to play a major role in battle—no more than three hulks can be shown to have contributed significantly to the Armada's fights—the fleet included thirty-one armed merchantmen of good quality from Spain's Atlantic ports, distributed, apart from those of the Indian guard, which sailed with the Castilian galleons, among the three squadrons of Biscay, Guipúzcoa, and Andalusia. Some of these were powerful ships, exceeding, on paper, the armament of all but the foremost galleons. The flagship of the Andalusian squadron, for instance, the Galician-built *Nuestra Señora del Rosario*, was among the ten most heavily armed ships of the Armada by number and weight of ordnance and store of shot and powder. The flagship of the Guipúzcoan squadron, the *Santa Ana*, had one great gun more than the *Rosario*—47 to 46—and was not far behind her in other respects, though the total weight of her broadside was probably rather smaller. Another *Santa Ana* was to have been the Biscayan flagship, but she proved unseaworthy before the campaign began in earnest, leaving the relatively modest *Gran Grín* of 28 guns to take her place at the head of the squadron and the *Gran Grifón* to fill the gap she left in the line of battle.

The great common strength of these Atlantic merchantmen lay, however, less in their armament than in their proven experience of northern waters. They were accustomed to the English Channel from their regular role in the Flanders wool trade. Such experience was useful not only as a test of the seaworthiness of the ships but also for the experience in pilotage of the men. 'Pilots practised and expert in the Flanders voyage' were one of the Armada's most acutely felt needs. Throughout the fleet, seventy-eight pilots, at most, were classified as 'practised' and there were a further four with knowledge not of the Channel specifically but of the north Atlantic: in the event, when the Armada had to find its long way home across that sea, this was perhaps to prove more useful. Though Sir George Carey found one captured pilot 'as perfect in our coasts as if he had been of Devon born', pilots who knew the English coast—and, even more, the treacherous shores of Ireland— were still harder to come by. As events proved, Juan Martínez de Recalde, who commanded the fleet's vice-flagship, was almost the only man aboard who could pilot a fleet safely into and out of a western Irish harbour by judgement rather than luck. The Spaniards made a remarkable effort to remedy this deficiency, printing and distributing throughout the fleet a

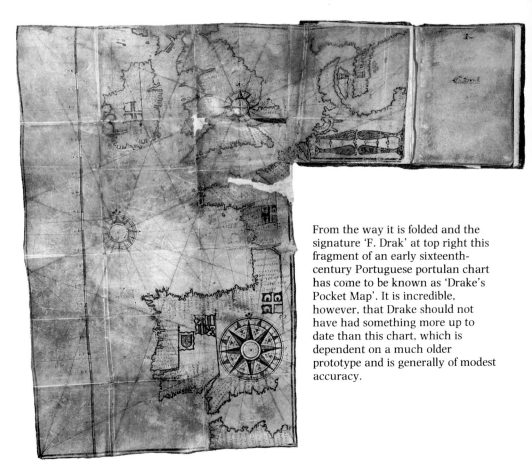

From the way it is folded and the signature 'F. Drak' at top right this fragment of an early sixteenth-century Portuguese portulan chart has come to be known as 'Drake's Pocket Map'. It is incredible, however, that Drake should not have had something more up to date than this chart, which is dependent on a much older prototype and is generally of modest accuracy.

ruttier of the Channel coasts and shores of the British Isles which was superior in its information to any other source that has survived from that time. It covered, among other things, the problems of pilotage into and out of the ports of England, soundings in the Bristol Channel, a summary of the coasts of Ireland, elaborate directions on the coasts of Normandy and Picardy, and a long discourse on the coastal tides.[12]

Largely untried in northern waters, but capable of arousing extremes of expectation and scorn, were the four galleasses contributed by Philip II's Kingdom of Naples, combining, in theory, the firepower of a galleon and the manœuvrability of a galley. Like galleys, they were at their best on a flat sea, where their oar-power could give them an advantage over ships that relied purely on sail—but the summer of 1588 was not such as to afford many calm days. For the rest, their armament was heavy and their construction surprisingly robust. They carried fifty great guns each—only the *Florencia*, among the galleons had more—probably including a high proportion of heavy guns, and were assigned the most exposed and critical positions on the wings of the Spanish battle-line. As we shall see, despite doubts of their adaptability to northern waters, they withstood English attack and the assaults of the weather with equal aplomb, save for one fatal weakness, which caused two of them to be wrecked and a third almost to follow them. It was with a disabled rudder that the flagship of the galleasses would come to grief in Calais; a broken rudder led to the destruction of the *Girona* on the coast of Ulster; and with the same defect the *Zúñiga* drifted uncontrollably about the Atlantic for more than a month before being grounded, but not wrecked, on the French coast (see below, pp. 189, 253, 264).

For its sheer heterogeneity, if not for its religious inspiration, the Armada would have deserved the title the Spaniards gave it of 'Catholic Fleet'. This makes it impossible to hazard a useful overall assessment of its fighting strength. There is no acceptable way of calculating the relative value of a Portuguese galleon, say, compared with a galleass, or of a lightly armed Castilian galleon with a well-armed merchantman, or of a heavily gunned but frail Levanter with a lightly gunned but robust Biscayan. It is not easy even to calculate the Armada's strength in the crudest terms—numbers of fighting ships. One rough-and-ready criterion is number of guns. Excluding the Biscayan *Santa Ana*, which left the fleet before the campaign began, there were sixty-five vessels, among those that certainly sailed, of more than twenty guns apiece, but eight of these were hulks carrying no heavy ordnance. When the Armada sailed in the formation of a convoy forty-three ships were considered good enough to form its outer husk. If, however, all the galleons and galleasses are counted, and the next ten most heavily armed ships in terms of weight of ordnance, all of which had guns of 18-pounder class and upwards aboard and were therefore at least as well armed as most galleons, the effective fighting strength can be reckoned at thirty-four ships; only twenty-nine can be shown to have made a major contribution to the fighting, either by virtue of being mentioned for gallantry in the contemporary relations or by virtue of evidence of battle damage exhibited at the time, and of these, four (the *Gran Grifón*, the hulks' vice-flagship *San Salvador*, the Biscayan *María Juan*, and the Guipúzcoan *Doncella*) would not qualify to be included on grounds of their armament. After the campaign was over, Medina Sidonia's official military adviser, Don Francisco de Bobadilla, who always tended to asperity of judgement and was particularly free with accusations of cowardice, complained that only some twenty ships of the fleet had performed effectively in battle. Though harsh, his assessment was perhaps not far from the truth. Drake, when he first beheld the Armada, put its fighting strength at not above fifty: on the evidence before us, that was not a bad guess; in practice, however, though that was roughly the right potential figure in terms of numbers of ships of sufficient size and firepower, the 'front line' may have been little more than half as strong. To assess its strength relative to that of the English, the ordnance has to be broken down into comparable categories, and the comparative speed and handling of the ships has to be taken into account. We can do this best in the context of the ships in action, in Chapter 6 (pp. 146–71) below.

The English logistic effort resembled that of Spain to a remarkable degree, save for its smaller scale and greater inefficiency. Spanish planning called for nearly 30,000 men at sea for three months. To English eyes— those of the purveyor Darrell—even 9,500 men 'to be furnished for one month will require a great mass of victual'. While the Armada was destined for operations nearly one thousand five hundred miles from the nearest friendly deep-water port, Darrell doubted whether the English would have sufficient shipping to shunt provisions around their own home waters. Yet England's 'great preparations and threatenings now burst in action upon the seas' could strike awe into Walsingham himself. They were strictly unparalleled in English experience. As Walsingham pointed out of the landward defences, 'there was reduced into bands and trained in maritime and inland counties, under captains and ensigns, 26,000 foot and horsemen: a thing never put into execution in any of her Majesty's predecessors' time'. On paper—as in Spain—the English plans were even grander, calling on a phantom army of 132,689 men, though it was admitted that many of these would be armed only with pikes and bills. The Lord Mayor of London was to raise and train 10,000; every county was given similar directions with absurdly optimistic targets. Some instances of previously unsuspected energy were called forth. In 1587–8 remarkable surveys of the coasts of Devon and Cornwall, Sussex, Kent, and the Cinque Ports, and Norfolk—and perhaps others which have not survived—were accomplished; in April 1588, the navy detained a Swedish ship full of masts and ordnance, presumed bound for Spain, 'to save the queen the expense of sending east for masts'. Meanwhile, the city of Norwich was paying its gunsmiths a three-shilling bonus to work at night. Creative sparks like these gleam against the background of chaos.[13]

Waiting in bed for an epileptic fit to strike, Walsingham wrote despairingly to Burghley as the Armada sailed out of Lisbon, 'I am sorry to see so great a danger hanging over this realm, so slightly regarded and so carelessly provided for. I would to God the enemy were no more careful to assail than we to defend.' He was in low spirits and inclined to

Musketeers and musicians of the London trained bands, the élite of England's militia, are shown in the funeral cortège of Sir Philip Sidney. Even on so solemn an occasion their bearing seems unmilitary, and though well uniformed and equipped, they display an extraordinary range of fashions in hose and garters.

exaggerate. Englishmen's awareness of the danger was real enough, but they responded to it slowly, like the peasants of Écija to Cervantes's pleas, because of long-ingrained, narrowly self-interested habits of thought and a traditional wariness of the demands of the crown. Every imaginable obstruction was flung in the way of the executive. To requisition ships even in the season of acute peril was a gruelling task. On 11 April 1588, for instance, the city of Exeter explained that her quota of ships could not be provided because 'by reason your Honours' letters are directed only to Exeter and Topsham, all the rest of the places and creeks belonging to the port of Exeter, whereunto the most number and best ships of the whole harbour doth belong, do allege that because they are not specially named in your Honours' letters, this service concerneth them not.' The following day King's Lynn provided a similar excuse, accusing neighbouring communities of being 'unwilling to be at any charge near the furnishing of a ship'. Thus the crown reaped the ill effects of local animosities. Orford and Dunwich quarrelled with Ipswich over the contribution of Woodbridge. Such disputes could have curiously lingering effects. As recently as 1929 Bideford and Barnstaple were still disputing the honour, which they had formerly vied to avoid, of having supplied five ships for the queen's fleet. In July 1588, the Privy Council, in evident exasperation, authorized the inhabitants of Lyme to use force against Axminster and other places around about 'wherein they are required to use their good endeavour, in respect that this charge is for a public defence wherein every man hath a like interest, that the said contribution may be gathered speedily for this public service so as their Lordships may no further be troubled about this matter'. At Foy and Low in Cornwall, a similar local solution was found, this time an amicable one: John Rusheley, Esquire, bore £500 of the £600 required for one ship and one pinnace to spare his recalcitrant fellow citizens. At Wells, on the other hand, the Council acted arbitrarily, seizing the ringleader of local resistance and taking bonds for his future compliance.

On the whole, it may be supposed, as Armada contributions were a form of taxation—royal encroachments on subjects' property—that resistance was concentrated at the upper levels of local society. Such, indeed, was the Privy Council's complaint: 'those that have houses, are of good ability and for the most part inhabit in the City refuse to contribute to the general charge of victualling'. In the case of Exeter, the Council acted to prevent victimization of the poor: as 'their Lordships were given to understand that they [*the authorities of Exeter*] purposed to collect towards their charge as well of the poorer sort as of those that were of good ability, they were required to spare and favour the poor so much as conveniently might be, and that the charge might be borne by the worthy, being best able to contribute towards the same.'[14]

These hints of social tension have to be understood against a background of genuine hardships. Economic distress, caused in trading communities by the commercial disruption of the war with Spain, was a frequently alleged ground of inability to pay for defence against the Armada. Ipswich claimed to be unable to pay for the ordnance charged against her. To the Privy Council's insistence that the cost be met by merchants who had profited from wartime reprisals against Spanish shipping, the municipal authorities 'for answer thereof affirm that they have thereby rather sustained loss than gain'. The port of Southampton wrote with a similar complaint in greater detail on 17 April: two large ships and a pinnace had been required of that town but its 'poor and insufficient number of inhabitants' would have been hard pressed to raise even the fourth part of the £500 this would have cost. War with Spain and consequent loss of trade was the main reason alleged—plausibly enough, in a place traditionally prominent in the Spanish trade and heavily concerned in its chief commodity, cloth. During the sixteen years of the Spanish embargo, the burghers said, there had been 'almost no other trade or traffic'; reprisals, as at Ipswich, had brought no relief— only a net loss of £4,000; native townspeople of substance had all gone away and there were virtually no gentlemen left. The community had already been taxed to its limit: a subsidy of £120 had been levied with difficulty; a recent charge of £250 for munitions 'remaineth dead and without profit to the town'; repairs to the seabanks and 'some little fortification' had exhausted the town's resources. To cap it all, the press had taken 110 mariners away to the fleet; so the town would have been unable to find men to sail the ships even 'if we were able to levy the charge among us'. In less strident, but equally self-righteous terms, the magistrates of East Bergholt also appealed to the economic effects of the Spanish embargo to explain their refusal to contribute, for the town's cloths

were best saleable in Spain and now through long want of vent into those parts, we find not only the stocks and wealth of the said inhabitants greatly decayed, but withal they, being very charitable and godly bent, are driven, out of their own purses, to see all the poor and needy artificers pertaining to the trade provided for sufficiently with meat, drink and clothes.[15]

The similarity of English and Spanish experience can be illustrated by comparison with Cervantes's efforts to levy supply in Andalusia or with Valladolid's response to the royal need of troops. Ordered on 7 November to allow recruiting in the city, the Council of Valladolid did not at first deign to reply. The members made openly defiant speeches, complaining of the cost of billeting, the disorders of the soldiery, and the 'inconvenience' caused by recruiting. One of the most outspoken councillors was Pedro López Enríquez, whom one would have taken for a perfect member of the knightly class. He had even translated Ariosto's great poem about the young Roland. But he would not let his chivalric tastes subvert his role as a patrician of Valladolid. He was, he declared, 'always disposed to serve' the king, but not to obey in this instance, as Valladolid had no men to spare and could not be expected to supply men without knowing the purpose for which they were intended; the plea of political secrecy meant as little to him as to 'civil rights' activists today. The truculence of Valladolid could only be overcome by the intrusion of a royal administrator, who temporarily took over the government of the city and arbitrarily admitted two captains to raise 250 men. Usually, however, local communities in Spain were content with a mere statement, short of a tooth-and-nail defence, of their rights. Logroño, for instance, declined the king's first request for recruitment to proceed on the grounds that the town had an ancient privilege against it; at the second time of asking, however, the captains were admitted, with a stipulation that this was to be without prejudice to the town's rights in future: henceforth the city would allow its sons to enlist, they promised, but would-be recruits would have to go to neighbouring districts to do so.[16]

In England too it was usually possible to come to some accommodation with objectors. The fathers of Kingston upon Hull, for instance, were

pleased to report on 20 April 1588 that since the time of their excuse that
the ships required by the crown were 'away upon their traffic' and
that impressment had depleted the seamen of the town, other ships and
mariners had returned from London and Newcastle, enabling the quota
of two large ships and one pinnace to be filled. There was, however,
sometimes a menacing edge to English recalcitrance, such as was not to
be found in Spain. When Colonel Dawtrey arrived to inspect the musters
of Hampshire—and 'very rawly furnished' he found them—he was dismayed to perceive that he was left to defend the entire Hampshire coast
with bill-men because the best levies had been poached for the garrisons
of Portsmouth and the Isle of Wight; though he recommended a new
muster, he warned that it would be dangerous to force the people to any
further expense. What the people of Hampshire—and, no doubt, many
other counties—feared most was that 'this new increase of Furniture or
Armour shall be a continual charge upon themselves and their posterity
hereafter': the crown's constant search for prerogative taxes had sown
deep distrust. In Southampton, 'the motion thereof to the people doth
cause them greatly to murmur and grieve thereat'. Mere delays could be
as deadly as denials. The Earl of Bath found it impossible to get a muster-
roll for Devon until the Spaniards were in sight off Portland and even
then it was only a 'very disordered scroll'.[17]

To both England and Spain, more may have been lost through peculation than through evasion of social obligation. Petty thievery was
inevitable from both stores and prizes, for, as Sir George Cary warned
Walsingham, 'watch and look never so narrowly, they will steal and
pilfer'. It was large-scale larceny, at the higher levels of society and of
command, that caused the greatest anxiety. Sutlers were generally
assumed to be dishonest. The rumour that a Lisbon baker had been
imprisoned for supplying the Armada with bread mixed with lime, the
accountant's warning to Philip II that unless the spoiled biscuit were
quickly discarded, 'someone will be sure to buy it up and mix it in with
the fresh biscuit, which will be enough to put paid to all the Armadas
afloat', and the English crown's long history of litigation against Elizabethan purveyors all illustrate the stereotype of the dishonest sutler. The
crews of the Guipúzcoan squadron complained of short quantities of wine
in the butts. Barnaby Rich's recruiting sergeant expected men to be
dunned: 'besides the exactions of the victuallers they shall be infected with
unwholesome and unreasonable provisions, oppressed by the provision
master, cheated by so many scraping officers that it makes you mad to
think of it'. The crown tried to protect its humble servants by rigorous
price controls and controls of quality which were, necessarily, more

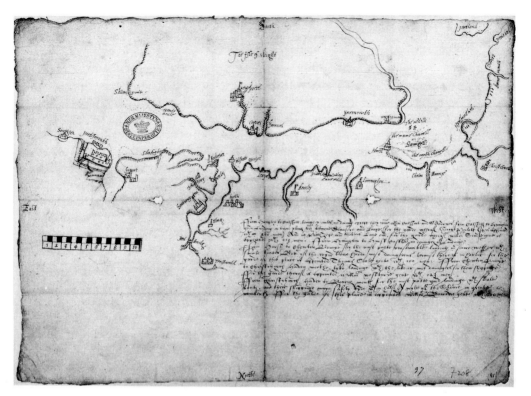

The threat of the Armada was an important source of stimulation to the surveying of the coasts of England. This map of the Hampshire coast and the Isle of Wight identifies possible landing-places for different sorts of craft and specifies the numbers of men available for defence. The document belongs to an early phase of English strategic planning, when it was envisaged that English forces would be dispersed to oppose Spanish landings, rather than concentrated for response in force.

imprecise. A threepenny supper in an inn for a soldier bound for Tilbury in 1588 had to include good wheaten bread and drink, boiled beef, mutton, veal or lamb and roast pork, beef, mutton, veal or lamb, or on fish days salt fish or ling, eggs, butter, buttered peas or beans, all 'competent and sufficient for the sustentation of his body'. Yet this was an outrageous counsel of perfection, for no such meal could have been provided at the price, even by the standards of the crown's own wholesale regulations. To judge from the prices laid down, the fleet seems actually to have got a fair deal from the purveyor-general, James Quarles. His price for beer, for instance, was well below that specified on land—a penny three-farthings a gallon instead of a halfpenny a quart. And, as he pointed out, he demanded nothing for 'biscuit bags, necessary lading charges, clerks' wages, surveyors' wages and all other incidents pertaining to the service'.[18]

The dishonest sutler was succeeded or abetted by the dishonest captain.

The system introduced into the English army, of making captains responsible for rations, was a notorious source of abuses, including inhuman exploitation by officers of their men. This was indicted by Shakespeare in his portrayal of Falstaff. The rumours that captains absconded from Tilbury with their men's pay seem unlikely to be true, not because captains were honest but because the pay was rarely forthcoming. Spanish captains seem to have been better—the tenor of army life gave Spanish officers less scope to despise their men—but a certain amount of thievery seems to have been a widely tolerated 'perk'. The accountant who supervised the Armada's finances on its return had no doubt that most captains had helped themselves to a pipe or two of wine. 'Captains are so many,' he wrote, 'and companies so few, companies having been bestowed so freely, that it is meet some man of authority should come hither with all speed to inquire ... because the amount of wages is so large.' And again, he recommended, 'It would be advisable to have a secret investigation of the notaries' books, taking them by surprise ... I think there has been a good deal of laxity about this, in consequence of the many deaths that have taken place.' Peculation was covered by a form of agreed taboo. When Juan de Cardona was ordered to investigate it, he responded with embarrassment and evasion. He could, he told the king, learn nothing about expenditure at Lisbon. He advised commissioning an expert 'who will deal with nothing else but find out quantities and qualities of these things, and I imagine that in both he will have great difficulty'. He had found that three bills drawn by Medina Sidonia at Corunna had not been duly countersigned: 'I send word to Your Majesty of this, for although the sum is a small one, it is an indication that whoever induced the Duke to sign them, if he was an officer, would go further still. I do what Your Majesty orders although these matters are not within my capacity, but it is better to keep silent when I do not know anything. God save Your Majesty.' In England, Hawkins feigned similar myopia in the face of comparable abuse. 'I never yet knew any profit by sea books,' he protested to Burghley after the Armada campaign,

nor know what a dead pay meaneth, as it hath been most injuriously and falsely informed. There are diets to the captains, dead shares to the officers, and such like accustomed pays to the officers, which are paid, and no more ... If I had any enemy, I would wish him no more harm than the course of my troublesome and painful life, but hereunto, and to God's good Providence, we are born.

The question of peculation evidently bedevilled the logistical effort on both sides; but its details must be left beneath the veil of collusive discretion with which contemporaries covered them.[19]

Two

The Image of the Enemy

IN these pages I want to pose Peterkin's problem of 'what they fought each other for', not on the level of the diplomatic gyrations which preceded the Armada and turned the English and Spanish crowns from traditional friends to deadly foes, but on the level of the mutual perceptions of the men who took part in the conflict. How did they see each other? And what interest did they have, or at least aver, in fighting each other? This may in part be a *question mal posée*, because, at least on the Spanish side, the war was waged by professionals, whose motivation depended little if at all on their mental picture of their adversary. English motivation may be thought to be adequately explained by the proximity of an invader, whatever his provenance—by 'foul scorn that Parma or Spain or any foreign prince should durst invade the borders of this realm'. Nevertheless, perceptions of the enemy are always influential in conflict: if not strictly on why, then at least on how one fights him, and it may prove worth while, without presupposing any necessary connection, to pursue the themes of perception and motivation in close proximity.

The task cannot usefully be approached except with some awareness of the type of men whose mental world is under scrutiny. Lower ranks make ineloquent warriors and it is hard to see them other than through the blurring mists of generalization. It is possible, however, to point to some differences, commonly supposed at the time, in the social composition of the Spanish and English forces during the Armada campaign. Philip II was in no doubt of the nature of the basic difference. He had a professional army at his command, an élite force in the social as well as the martial sense, whereas 'that of the enemy is raw, plebeian, disorderly and bereft of military discipline'. This remark may have been recklessly overdrawn, but it was no baseless caricature. The Spanish army was indeed an honourable occupation, in which a gentleman might start in the ranks and spend his active life in the service. For a penurious hidalgo, the average expectation—a captaincy after nine years' service—was by no means contemptible; and the army was something of a *carrière ouverte aux talents*. A captaincy was the goal—the picaresque memoirist Alonso de Contreras, whose pretensions to gentility were of the shakiest, was

forever soliciting one and got one at last—and even the humblest ranker
could be said to have had the hope of one 'in his knapsack'. Soldiering
was a career with a clear structure of ascent. Moreover, it could command
prestige, always the aim of the upwardly mobile. 'Sir, I've no use for
money,' said Alonso de Contreras. 'If there's anything I particularly want,
it's renown I seek, not cash.' Recruitment was voluntary: it was at free-
willed, independent men, 'the men who win victories', as the Duke of
Alba remarked, that captains aimed when they set up their colours. When
the recruit took the oath to obey orders and renounce desertion and
mutiny, he retained the dignity, at least, of an independent way of life. It
is probably fair to say that a Spanish soldier's self-image was derived from
romances of chivalry and that his notion of his place in the regiment was
one of knightly companionage. All this made, no doubt, for a better army
than was possible in the English system, which imposed military service
as a social obligation or bestowed it as a form of relief for the dregs of
society. Length of service is directly related both to technical proficiency
and professional pride: only a handful of English soldiers had any chance
to acquire those virtues. It is hard to imagine a Spanish captain saying of
his men what Falstaff said of his: that they were 'food for powder' and
'good enough to toss'. Awareness of Spanish superiority, indeed, may
have underlain Shakespeare's satire. In other attacks on the deficiencies
of English recruiting traditions the implied comparison with Spain is even
stronger. Barnaby Rich's Captain Pill, for instance, sees what he thinks is
a vagrant being dragged off to be hanged. On closer inspection, he finds
an idle rogue pressed for a soldier. He confronts the recruiting sergeant
with a demand that he enlist better men. The sergeant is outraged. Pill,
he perceives, would like to have honest householders placed in the ranks,
'of wealth and ability to live at home, such as your captains might chop
and change and make merchandise of ... But God defend that any man
of honest reputation should be levied just to be extorted.'[1]

It may be, of course, that the perceived difference between the rival
armies was greater than the real one. Spanish recruiting standards seem
to have been in decline in some respects. The actual behaviour of Spanish
troops decayed with extraordinary rapidity in the years after the Armada.
The success of Spanish arms in Europe in the sixteenth century was not
produced solely by valour in the field. Most warfare was positional, and
what counted was the power to keep armies in the field. Spanish armies
were superior because they stayed together for longer than those of their
enemies before mutinying or melting away, victims of the wasting disease
that was endemic in early modern warfare. Before the Armada, the Army
of Flanders had experienced its share of mutinies or 'furies'—as the

ravages of licentious soldiery were called when the phrenesis of indiscipline came over them. But they were conspicuous, in part, precisely for their rarity. After 1590 there were more than forty serious mutinies in twenty years. This was somewhat worse than what might be called the 'going rate'. Spanish arms had lost their moral advantage, and the material advantage that accrued from it.

The incidence of mutiny was governed, of course, by pay, conditions, and fortunes of war. The greatest of these, perhaps, was pay, or material reward more generally. Against Alonso de Contreras's repudiation of mercenary motives, we must set the well-known proverb recorded in *Don Quixote*:

> To the wars I am carried by nothing but need.
> If I had any money I should not go indeed.

In the fiction—and even, in Contreras's case, the autobiography—of the next generation, the Spanish soldier has few noble traits: he will become a picaroon, culled from the gutter, to which, far from being permanently elevated by military service, he will ultimately return. If he is willing to endure army discipline, it is only for the sake of the plunder and pay, or simply, more often than not, to escape from the consequences of crime.[2]

Meanwhile, of course, ordinary men had honourable uses for cash, which, in the right circumstances, were worth a mortal risk. One of the most poignant memorials to have survived from the Armada is a soldier's letter home, addressed to a friend from Corunna about a week before the fleet sailed, with a message for his wife, which declares his motive for going on the expedition:

To Ana de la Cueva Your Grace will say that she should commend me to God, for if there is one thing for which I wish to go on living it is to return to her and to provide for her as by rights I ought, and if God gives me life, when this expedition to England is over, which will be very soon, I shall go back to my house again with whatever God will be pleased to give me in order to have some peace. For it was for this that I came on this expedition, not because I am any lover of wandering away from home and God knows that all I want is to be there again in peace and with some little rest.

Medina Sidonia harboured no illusions about his men's need of pay to keep them loyal. His most forceful plea for wages and cash in hand for other necessities was written on 28 April 1588 when he was expecting to sail imminently—perhaps without the longed-for pay chests. 'The soldiers and sailors,' he wrote,

being so poor, and in such need, as they are, and being made to embark without receiving more than two pays, which will not give them enough for their shoes,

Your Majesty will readily believe that they will go in great discontent, and even more so if it is understood that the Armada has no money on board, for this must dishearten them greatly, and with men who are so discontented and ill paid, it makes it very difficult to do any good service, and therefore I beseech Your Majesty to consider this, as the time is so far forward, and even if it means using the postal service and despatch riders on horseback, it would be wise to send forthwith a big consignment of cash.[3]

As well as those who enlisted out of necessity or despair, both sides had a large complement of gentlemen-adventurers. The *jeunesse dorée* of Spain flocked to the fleet, though it seems there were more captains commissioned than companies raised. Gentlemen-adventurers attended on their own or with their personal retinues, at their own expense unless they already had, or now secured, a royal pension. The Marquess of Santa Cruz had included estimates of the cost of entertaining noble fellow travellers at his table in his accounts. This could mount up formidably. At a time when a soldier's pay was three *escudos* a month, it cost 500 *escudos* to entertain a commander's entourage for the same period. Santa Cruz thought noble amateurs a nuisance; 'each one of them', he complained to the king, 'thought to build bedding and lodging for himself on board and this was a great inconvenience, because they took up space needed for fighting and got in the way of the seamen in sailing the ships'. Medina Sidonia confessed he could do nothing with such men. He wrote to the king on 2 April 1588:

Here are so many knights, soldiers and suchlike private volunteers whom I have to satisfy, and all with fine records and nowhere to accommodate them, that I do not know how to employ them, unless Your Majesty should be pleased to assign the wherewithal for their entertainment, for although I do not know them I am told they are good fighters and useful.

An English gentleman-adventurer cut much the same figure as his Spanish counterpart and, in his commander-in-chief, inspired similar *ennui*. When aristocratic volunteers began to gather at Plymouth in February 1587, Howard heartily wished them back at court. Between the lines of his letter to Walsingham, the swagger and ostentation they evinced and the embarrassment they aroused can be detected: there was nothing to do during the period of phoney war, he complained, and the gentlemen wasted their time and money on idleness and conspicuous consumption. 'I am more sorry for the noblemen than any ways for myself; for I would have them save, to spend when need shall be. I do assure you they live here bountifully and it will be hard finding of such noblemen as these be, so well affected to this service and that will love the sea so well as they

One of the most poignant artefacts yielded by the Armada wrecks is this ring from the *Girona*, inscribed 'Madame de Champagney MDXXIIII', and worn—it must be presumed—as a family heirloom by her grandson Don Tomás de Perrenoto, who is known from other evidence to have perished on board. He was Cardinal Granvelle's nephew, one of the most promising and best-connected of the gentlemen-adventurers who died on the expedition.

do.' All complaints against gentlemen-adventurers are like this, mixing annoyance at their habits with acknowledgement of their prowess. In any case, the social convention that committed the nobility to war was too ancient and too strong to be set aside, even in an age of increasing professionalism in combat. The sea, moreover, had a surprisingly large role in chivalric literature as a proper field of great deeds, especially in Spain, where one of the most influential of chivalric classics, the *El Vitorial* of Count Pero Niño, had a sea-borne setting, and where in innumerable poems, romances, and chronicles of chivalric deeds, the waves were ridden like jennets. This tradition made the sea an irresistibly suitable medium for displays of martial virtue by the knightly class. And although the old-fashioned chivalric virtue of good vassalage—service to one's lord as a sufficient ground to go to war—was no longer much mentioned, there were still gentlemen-adventurers who prized it. Sir Horatio Pallavicino may be selected to speak for them, even though, as a naturalized, Genoese-born subject of Elizabeth's, he had a peculiar motive for stressing the importance of loyalty. 'The greatness of my zeal,' he wrote to Walsingham in explanation of his decision to join the campaign,

which desireth to be amongst those who do fight for Her Majesty's service and for the defence of her kingdom, doth constrain me, with an honourable company, to depart as this night toward Portsmouth, there to embark and join the Lord High Admiral, where I hope to be present in the battle, and thereby be a partaker in the victory or to win an honourable death, thus to testify to the whole world my fidelity to Her Majesty.[4]

The common soldiery on both sides, then, were perhaps less unalike than most of the sources suggest; the gentlemen-adventurers were evidently very similar. The third major component of the forces were the seamen themselves, who were recruited on both sides in much the same way, from much the same class of men. Both navies relied on a permanent professional service to man the relatively small royal fleets, supplemented at need by the impressment of practised mariners. If any difference of kind was perceived at the time, it seems to have been by the English, whose attitude to the respective merits of the two countries' navies was as vainglorious as that of the Spaniards to their respective armies. In a common English opinion, the Spanish seamen lacked the social cachet and dedication of those of England. Burghley's pseudonymous propaganda

Playing-cards were a popular medium for anti-Catholic propaganda in the 1680s, when the series known as the Armada pack was made: it is evidence of the part played by memories of the Armada in the anti-popery of the late seventeenth century. The chivalric gesture by which battle was preceded—the exchange of challenge and defiance—is recalled rather with pride than derision.

The English Pinnace caled ye Defiance sent from the Admirall, and by a great Shot Challinging the Spaniards to Fight. the 2j of Iuly 1588.

tract, the *Mendoza Letter*, for instance, claims with little justification that the great strength of the English at sea lay in the many ships manned by volunteers from port towns who served without wages, whereas the Spaniards were thought from English intelligence reports to be 'taking up all the men and mariners in Biscay, whether they will or no; but to content them [the king] imprests them with twenty or more ducats beforehand. Also he takes up all fishermen, old and young.' The idea that a free force of citizen volunteers was superior to a 'slave' or mercenary force was one which humanists claimed to have learnt from classical antiquity; it was Machiavelli's only respectable lesson. The contrast between the two navies in English eyes was therefore probably produced by the contemplation of a literary tradition rather than by scrutiny of the real situation. It was remarkably like the contrast which Spaniards professed to see between the two armies. Broadly speaking, therefore, the Armada conflict was a struggle of like with like. What ground of violent antipathy, what fatal objection, could be raised or exploited against the foe?[5]

Religion was the most obvious source of an ideology of enmity. Religious fervour is hard to sustain, but, in short spurts, easy to encourage. No one doubted in the sixteenth century that violent means could be hallowed by religious ends. Spanish seamen habitually thought of their adversaries as confessional foes. The shanties Eugenio de Sálazar heard on his way to Santo Domingo in 1574 assume that every enemy, of whatever provenance, is a Moor. On the haul of the rope, the men were exhorted to 'maintain the faith' and to 'confound and kill ... Muslims, pagans, and Saracens'. On the slack, they would shout a cry of assent:

> Oh, they deny—the Holy Faith,
> Oh, Holy Faith—the Roman Faith!
> Oh from Rome—does mercy flood,
> Oh from Peter—great and good.

If the anchor cable was a long one, or a sail took long to hoist, these exchanges could be considerably protracted, growing more secular in tone and content as time went on and improvisation grew more desperate. By the end of the shanty, thoughts of faith and war might yield to the seamen's more worldly concerns, like wind and sun, love and youth: *O levante—se leva el sol / O ponente—resplandor / fantineta—viva lli amor / o joven home—gauditor!*

> From the east—the sun arose.
> In the west—the sunset glows.
> Hey, my girl—may love last long!
> Rejoice, my lad—while you are young.

And so the shanties proceeded, with increasingly ribald, wild paganism set to the quasi-liturgical alternation of versicle and response. But the importance of the basic appeal to religious inspiration, and the way in which a confessional explanation and justification of conflict are encouraged by what one might call the 'prescribed' chants seem undeniable. Spanish Armada propaganda exploited—or tried to exploit—this tradition. The men of the fleet were exhorted to save 'the tender children, who, suckled on the poison of heresy, are doomed to perdition unless deliverance reaches them betimes'. In Lope de Vega's imagination, the masts of the Armada were 'roods of faith'; the ships, in their gay colours and streaming banners, wore the crimson vestments of a martyr's feast.[6]

There were particular reasons for seeing the invasion of England in this sort of way. It is a matter of doubt whether the English people or their queen were sincerely committed to Protestantism. Elizabeth's own devotional tastes were conservative; most Englishmen, probably, still knew little of the Reformation and cared less. The new religion had been imposed from above, and it was at relatively high levels of power in church and state that zeal for it was concentrated. Yet the Elizabethan regime depended on a Protestant élite at home and Protestant allies abroad. The English generally had thus acquired notoriety not only as heretics themselves—'Lutherans and chickens' in the terms of the insult hurled from the decks of the Armada—but fautors of heresy, in France, in Scotland, and, most ominously from the Spanish point of view, in the king of Spain's own dominions in the Netherlands. The Dutch war had not started as a war of religion, but had become one. As participants strove to clarify their own motives and draw the lines of battle in their favour, religious differences came to assume enhanced importance; dogmatic positions became increasingly sharply defined. Elizabeth flung English volunteers into the conflict and regular troops from 1585. The war had the same effect on them as on the indigenous participants; the *odium theologicum* was infectious. The Catholic Sir William Stanley marched his men over to the Spanish side, declaring that he had ceased to serve the devil and had turned to the service of God.

As well as the Netherlands, the other great theatre in which Spaniards and Englishmen clashed in the 1580s tended also to highlight religious differences. English merchants were appearing in increasing numbers in the tribunals of the Inquisition. Few of them were ardent Protestants; but they came from a land which had now been without any but a clandestine Catholic ministry for a generation. Any ignorance of the Roman faith, or ill-assimilated influence of that of the reformers, was bound to be harshly judged in Las Palmas or Seville. The effect was two-edged: in Spain,

The faded fragment of a pennant from the *San Mateo*, grounded near Nieuport after the battle of 8 August. The figure-drawing of the crucified Christ is reminiscent of the Hispano-Flemish style of the late fifteenth century. Naval banners may well have reflected conservative artistic coventions; on the other hand, this example may already have been very old at the time of the Armada.

the notion that Englishmen were irremediably tainted with heresy was strengthened by the foregone conclusions of the Inquisition. In England, Foxe's efforts to inspire horror of the Inquisition as an 'engine of tyranny' were confirmed.

More than any other single focus of attention, however, it was probably the images of the rival monarchs, Philip and Elizabeth, that fixed the hostile perceptions of the 1580s on religious lines. Elizabeth had every reason, apart from true zeal, for presenting herself, and having herself presented, as a Protestant saviour. The insecure throne she ascended in 1558 needed firm supporters and none were more serviceable or more determined than the godly élite—those who had drawn inspiration from the messages of Luther and Calvin or profit from the Reformation of Henry VIII; those who had benefited from the largesse of Edward VI, and who had survived or been exiled under Mary. Elizabeth permitted herself to become identified with the cause that engaged their energies. Thus she became Deborah in Israel, the custodian of the Protestant Phoenix, the godly champion against the Romish Antichrist. In her roles as Astraea— maiden goddess of a 'golden age'—and Virgin Queen she seemed to mock the Mother of God. Such Protestant allusions abounded in official portraits and panegyrics, court entertainments and displays. It made for successful propaganda, and the image was accepted at home and abroad. The reality, of course, was quite distinct. Elizabeth seems to have had little use for theology except as an intellectual game, but, when it came to devotion, she knew what she liked and her tastes were distinctly old-fashioned. She worshipped in the well-cocooned splendour of the Chapel Royal, where traditional texts were still sung in Latin or in complex polyphonic settings which made only slight concession to the Protestant preoccupation with the distinctly audible word. Many of them were composed by the Catholic Byrd, whom royal patronage protected from persecution. She found clerical marriage distasteful and would not tolerate presbytery. But most onlookers saw only the Protestant image, not the reluctance with which it was espoused, only the dweller in Zion, not the hesitancy of her ascent.

Philip, with only a little more justification, sought to project in the Catholic world a near mirror-image of Elizabeth's in the Protestant. In his youth he had been quite differently portrayed, as a prince of the pagan Renaissance, in gilded armour, or even, perhaps, playing the organ in serenade to the goddess of love. In the 1560s, however, that image was discarded, unbecoming as it was to a great monarch with grave global responsibilities in the age of the Counter-Reformation. He seems to have undergone something like a religious conversion at about the end of the decade—connected perhaps with the tragic end of his son and heir, Don Carlos, who died, imprisoned by his father's command, after showing symptoms of violent madness. At about the same time, negotiations to form the Holy League with Venice and the pope for a sea-borne crusade against the Turk helped to stimulate a mood of Catholic triumphalism at Philip's court, symbolized by Titian's great allegory of 'Spain Coming to the Aid of Religion'. This atmosphere determined the reworking of the royal image. Philip became the epitype of Catholic piety, the sword of the church. When El Greco depicted the Holy League in 1576, Philip's image was stripped of every Renaissance adornment, and the king was left, kneeling, clad in penitential black, at prayer in the very maw of hell, while the hosts of heaven brandished the shining name of Jesus above his head. Like Elizabeth's Protestant image, Philip's Catholic one concealed a certain irony. His own conflicts with the pope often trembled on the brink of violence; far from embodying the spirit of unremitting crusade, he devoted a great deal of effort, in the interests of 'prudence', to restraining the fervour of others. In particular, he had been decisive in protecting Elizabeth from becoming the victim of Catholic irredentism early in her reign. By the time he changed his own mind, her throne was less vulnerable than ever.

At the highest levels of command, the Spanish expedition was seen as something very close to a crusade—that is, a war hallowed by remission at least of the temporal consequences of the participants' sins. No formal papal indulgences were issued for the Armada, though Medina Sidonia

El Greco offered the *Allegory of the Holy League* to Philip II as a sample of work in a successful bid for patronage at the time of the decoration of the Escorial. We can therefore be sure that it reflects the king's self-image of the mid-1570s with some fidelity. In contrast to the gilded youth painted by Titian a quarter of a century before, we are shown the mature Philip, clad as a penitent, worshipping the name of Christ, accompanied by his Catholic allies, and kneeling, precariously but serenely, at the very mouth of hell.

Titian's *Venus and the Organist* reflects the ambience of Philip II's *jeunesse dorée*. The values for which it speaks—pagan classicism, secular art, luxury, and erotic sophistication—contrast with the puritanical sobriety evinced by Philip with increasing severity from the 1560s. The fanaticism that launched the Armada was the more intense, perhaps, from being inspired by something like a conversion experience.

strove to wrest spiritual concessions of some sort from Rome. The king informed him on 22 April 1588 that the 'faculties' he had sought from the papal nuncio had not been forthcoming; the pope had, however, decreed a jubilee and the proclamation was 'to be carried aboard the Armada and published and the jubilee may be gained at any time, as best may be, for the purpose for which it has been decreed is to help the success of the expedition on which you are embarked.' For the rest, he had to be content with a general benediction upon the enterprise by the Archbishop of Lisbon and a standard embroidered with the Crucifixion on one side and Our Lady on the other, raised from the high altar of Lisbon cathedral with the legend upon it: *Exsurge, Domine, et vindica causam tuam.* Medina Sidonia may have been disappointed with this outcome—certainly he seems to have set greater store by the prospect of an indulgence than did his more world-weary and less ultramontane master; but if he had any complaint at the imperfect fervour with which the church upheld the sanctity of his mission, he made up for it by calling the invisible church to his aid. His draft exhortation to his men included a virtual litany of English martyrs.

The saints of heaven will go in company with us, and especially the holy patrons of Spain and, indeed, those of England, who are persecuted by the heretics and cry aloud to God for vengeance, will come out to join and help us, and those who gave their lives to establish the holy faith in that land and washed it with their blood. We shall find waiting for us there the help of the Blessed John Fisher, Cardinal-bishop of Rochester, of Thomas More, John Forrest and innumerable saintly Carthusians, Franciscans and other religious, whose blood was cruelly spilt by King Henry and who call on God to avenge them in the land in which they died. There shall we be helped by Edmund Campion, Ralph Sherwin, Alexander Briant, Thomas Cotton and many other reverend priests, servants of Our Lord, whom Elizabeth has torn in pieces with ferocious cruelty and nicely calculated tortures. With us will be the blessed and guiltless Mary, Queen of

Titian's *Spain Coming to the Aid of Religion* of the early 1570s expresses the spirit of militant Catholic dogmatism that was invoked to justify Philip II's wars from that time onwards. The allusion to the Turkish menace in the middle background was probably painted in to stimulate the solidarity of the Holy League at the time of the battle of Lepanto (1572); but the painting as a whole was intended to be a general-purpose image, which was reproduced in various versions with different background allusions to suit the occasion. Religion is shown wilting and downcast; her arms, cross, and chalice have fallen from her grasp; she is threatened by serpents. Only the timely appearance of Spain can rescue her.

Scotland, who, coming fresh from her sacrifice, displays ample and resounding evidence of the cruelty and impiety of that Elizabeth and utters her plaints against her. There too we are called by the laments of innumerable Catholics in prison, the cries of widows who lost their husbands for the faith, the weeping of virgins who were forced to yield their lives rather than destroy their souls, the tender children, suckled on the poison of heresy, who are doomed to perdition unless deliverance reaches them betimes; and finally myriads of workers, citizens, knights, nobles and clergymen, and all ranks of Catholics, who are oppressed and downtrodden by the heretics and are looking to us for their liberation.[7]

Even more than as a crusade, the expedition was seen by Spaniards as a sort of propitiatory sacrifice offered to God. English complaints of Spanish superstition, though made on other grounds, were thus strictly justifiable, for while penance is a proper Christian ground of sacrifice, propitiation is not. The propitiatory theme is at its clearest in an exchange between the king and Medina Sidonia after the storm which confined the Armada to Corunna in June 1588. Medina Sidonia—ever alert for arguments against the expedition—suggested that the storm was a clear sign of divine disfavour, in view of which the voyage should be cancelled or deferred. His master's unwavering reply contains this revealing paragraph:

If this war were unjust, the storm might be understood as a sign that Our Lord's will were to desist from offending Him; but since it is so just as it is, we must not believe that He would withhold His protection from it, but rather favour it beyond our desires. But if it should prove to be His will to reserve for Himself the punishment of those people, then, accordingly, His Majesty's intention will have been fulfilled on his part by having striven to serve God with all the power which He has granted him in defence of His cause.

The king's thinking was evidently muddled—if he acknowledged that God might 'reserve for Himself' the punishment of the English, how could he be sure that the Armada was in conformity with divine will? Yet the conclusion seems irresistible: Philip was reconciled in advance to the possibility of defeat, because it would magnify his sacrifice. The same mixture of propitiatory and penitential language characterized most of his letters on the subject. Again he wrote, if the enterprise were abandoned,

The enemies of the Catholic religion would interpret the damage inflicted by the storm as authority for their heresies, twisting in their favour God's tolerance, which was perhaps intended to chastise our sinfulness, or perhaps to enhance His glory at the time of our future successes. No step must therefore be taken by us to interrupt the course of the Divine purpose.

This line of thought provided a ready source of consolation in failure. For a more or less orthodox and thoroughly representative explanation of the outcome of the Armada as a gesture of Providence, wonderful, but not inscrutable, in His ways, one can hardly do better than cite Bernardino de Mendoza's letter of condolence to his king, written on 2 November 1588, when the full extent of the débâcle was apparent even to that optimistic ambassador. Philip might have been pardoned for disregarding Mendoza's correspondence by that time: the ambassador's agents had plied him with false information throughout the campaign, so that English propaganda could mock him as 'not de Mendoza but rather de Mendacia'. Yet the king retained his respect for his envoy and read the letter approvingly, underlining the pertinent quotation from St Augustine. 'Although I already had news', Mendoza wrote,

of the arrival of the Duke of Medina Sidonia with the greater part of the Armada at Santander, I humbly thank Your Majesty for informing me of it. With regard to the failure to achieve the desired goal, I can only repeat to Your Majesty what St Gregory says in one of his epistles: Adversitas ... probatio est virtutis, non judicium reprobationis, and as an example of this he instances the terrible torment suffered by St Paul when he landed at Malta on his way to preach the faith of Christ in Italy, and the King St Louis, whom God had elected as His own, suffered no slight adversity on his expedition to the Holy Sepulchre, notwithstanding his own saintliness. A single sin—much more the multitude of sins we men commit every day—forms, so to speak, a barrier between ourselves and God. And even when what we pray for is good and just, He does not grant it easily, in order to test our constancy and try our zeal in His service, and to lead us to correct our faults. We therefore may hope from His infinite goodness and clemency, that He will accord to Your Majesty success in the enterprise in proportion to your holy zeal in undertaking it, and that He has delayed success, in order that when it comes, it may evidently be the gift of His hand, and redound to His greater glory. For it will be seen that our Lord always precedes the greatest successes and victories by drawbacks and difficulties, and leads His chosen ones in His own way.[8]

Unless one discounts, on grounds of calculation and insincerity, everything he ever said on the subject, one must accept that Philip II saw history as providentially regulated and the Armada as part of God's work. During the approach to the campaign, he spent two or three hours every day 'on his knees, before the sacrament'. Juan de Vitoria was in no doubt that the higher figure was correct. The king ordered 'continual' prayers at court and urged them in the fleet. When the expedition was over and the scale of the disaster known, he wrote to the bishops of his realm, counting himself 'well served' by a programme of prayer, which, if it had not brought victory, must be presumed to have limited the scope of the catastrophe, and ordering a final solemn mass of thanksgiving in every church.

In his first instructions to Medina Sidonia he stressed, as the first priority, the need to preserve the participants in the expedition in a state of grace, communicated, confessed, and alert against sin, as in the old crusading tradition, in pursuit of divine favour and hence of victory; 'because victories are gifts of God . . . since you bear a cause which is so much His own that He therefore makes promise of His favour and help, if not foregone by sinfulness, you must take great care that, in the Armada, sins are avoided.' As he added soon after, prayer for the Armada's success, though essential, was insufficient in itself:

It is convenient and necessary, in addition, that the men aboard the Armada, for their part, strive to live in a Christian manner and avoid oaths and blasphemies and other vices, which are so offensive to Our Lord and because of which He sometimes permits that matters, albeit undertaken for His service, should not attain the end desired for them.

These prescriptions seem hopelessly unrealistic. The men of the Armada behaved like any other mass of soldiers and sailors, crowded together in taxing conditions, under a demanding strain. Yet Philip's idealism seems sincere—sincerely intended by the king, sincerely espoused by Medina Sidonia, who banned swearing and dicing and instituted searches of the fleet to root out whores. Such were the standards of a holy war in a 'godly' age.[9]

The Armada was launched, as Juan Martínez de Recalde admitted, 'in the confident hope of a miracle'. Though Philip II was inclined from time to time to doubt the fighting calibre of the English fleet—misled, it seems, by surprise at the small number of troops the English carried—and his ambassador in Paris, Bernardino de Mendoza, hastened to write the English vessels off as worm-eaten and ill-manned, the general atmosphere in which the fleet was gathered was one of apprehension at the immensity of the task it faced. The strength of the English was such, and the difficulties of operating in English waters so awesome, that without divine aid the expedition would have been unthinkable. The Venetian Lippomani exactly captured the mood at the Spanish court when he reported to his government that 'Everyone hopes that the greater the difficulties, humanly speaking, the greater will be the favour of God'. Philip's two chief commanders, Parma and Medina Sidonia, repeatedly warned him that the proposed operation was strictly impossible in human terms. Every practical objection they uttered was answered by the king with an appeal to Providence: 'this is a matter', he told Medina Sidonia, 'guided by His hand

and He will help you.' And when it seemed as though the Armada might never be able to undertake its mission, in July 1588, the king sought the same note of reassurance: 'I hope that Our Lord will turn all these hardships, which we are encountering at the beginning, into His greater glory at the end.'[10]

The English too, at least in official sources, seemed inclined to see the encounter with the Armada as a war of religion. It was, in the opinion of the Council, a typical device of 'the incessant malice of the enemies of the Gospel' who 'sought by all ways possible the overthrow and subversion of such as professed the same'. To the Lords and Commons in Parliament 'the thraldom of Romish tyranny' was an ever-present threat to 'the sincere and true religion of Almighty God'. By trying to project an image— by no means fully justified—of himself as the sword of the faith and of Spain as the helpmeet of religion, Philip II played into the Protestant propagandists' hands. They could present England as a 'beleaguered isle', ringed by Romish adversaries, whose ringleader was Spain. Protestant zeal animates the individual professions of motivation of godly participants in the Armada campaign. John Hawkins was typically robust:

Having of long time seen the malicious practices of the papists combined generally throughout Christendom to alter the government of this realm and to bring it to papistry, and consequently to servitude, poverty and slavery, I have a good will from time to time to do and set forward something as I could have credit to impeach their purpose.

A curious measure of the extent to which the two sides in the conflict shared the same outlook is provided by the way Protestant Englishmen were as inclined as Catholic Spaniards to see the fight as a trial before the God of battles. 'In open and lawful wars,' wrote Hawkins to Walsingham in February 1587, 'God will help us, for we defend the chief cause, our religion, God's own cause; for if we would leave our profession and turn to serve Baal (as God forbid, and rather to die a thousand deaths), we might have peace, but not with God.' Those words and sentiments, down to the implied analogy with the sacred history of the Jews, would have been equally typical in a Spanish source. The power of prayer was at least as important to the English as to their foes. Indeed, it was—unjustly—for neglect of prayer that the pamphleteer Daniel Archdeacon impugned the 'Spanish Nimrod' with all the self-righteousness of a practitioner of a religion of grace, attacking the pride of those who confided in works. It was, in his view, 'a continual practice of the wicked and ungodly, not only neglecting the almighty to trust in their own might: but relying on themselves and their own power'. The English clergy were mobilized for prayer, as well as taxed for more material contributions to the war effort.[11]

Nothing so clearly demonstrates the religious understanding of the conflict in England as the treatment of England's own Catholic recusants at the time. The queen and many of her entourage were anxious to display their confidence in Catholic loyalty. The *Letter to Mendoza*—purportedly written by a Catholic Englishman but really the work of Lord Burghley himself—is largely devoted to making this point. English Catholics, the work claims, 'begin to stagger in their minds, and to conceive that this way of reformation intended by the Pope's Holiness is not allowable in the sight of God', and the Spaniards are accused of unchristian conduct in embarking upon a forcible conversion of the English. Allen's propaganda against Elizabeth, Burghley claims, far from igniting Catholic rebellion, had united all the people in defence of their queen and country. Pietro Ubaldino—mouthpiece successively of Howard and Drake—was even more emphatic. It was, he claimed in obvious allusion to the memory of Philip's time as king of England, 'easier to find flocks of white crows than one Englishman (and let him believe what he will about religion) who loves a foreigner, either as a master or companion in his own home, even if a benefactor'. This apparent confidence was not borne out by the way the recusants were hounded and victimized. First, they were disqualified from training militia bands. Then their arms were seized and sold; then they were the subject of a campaign of fear and rumour-mongering. Bishop Cowper of Winchester unmasked two hundred Catholics living suspiciously 'in a little corner by the coast'. In every county, the representatives of the crown were ordered to watch the prominent recusants and keep them from each other. Finally they were interned in their hundreds, though many of them would willingly have served against the invader—like the German Jews ironically interned in Britain in the last war. It was a grim rejoinder to Sir Thomas Tresham's plea to the queen, 'Let not us, your Catholic native English and obedient subjects, stand in more peril for frequenting the Blessed Sacrament and exercising the Catholic religion (and that most secretly) than do the Catholic subjects to the Turk publicly.' The recusants were overwhelmingly loyal; the case they made on their own behalf was strong. Yet Elizabeth and her counsellors took the view that in a war of religion 'he that is not for us is against us'.[12]

All the evidence of confessionally self-conscious adversaries, with positions dogmatically defined, may, however, be misleading. Neither the Reformation in England nor the Counter-Reformation in Spain had yet had time to percolate through society. Despite the Reformation, Catholic and Protestant Europe still both belonged to western Christendom and shared a common culture. Though there were few fully fledged *politiques* in Spain or England—for France was the first land of inter-confessional coexistence—sensitive men could discern a considerable overlap in doctrine and devotion. The Yorkshire divine, Edmund Bunny, produced an Anglican edition of the *Christian Directory* of the Jesuit controversialist Fr. Robert Persons, who was a self-avowed 'subject' of Philip II. Bunny undertook the work in 'an eirenical spirit, hoping that well-willers would appreciate common ground' and decried 'books of controversy engendering inordinate heat'. An English translator of the *Vanidad del mundo* of the Spanish Franciscan Fray Diego de Estella declared that 'though the author be a papist' yet the work contained much wholesome and useful doctrine. Spanish devotional writings remained popular in England throughout Elizabeth's reign—considerably more so than in any other part of the Protestant world. Spanish culture was widely admired and imitated, at least until the Armada dispelled what respect remained. The English élite studied the Castilian language. The queen herself could speak Spanish to her Spanish horse. Sir Philip Sidney, who died fighting the Spaniards at Zutphen in 1586, knew his foes' literature intimately, and his own works are saturated in its influence. Dress followed the Spanish fashion.[13]

And if Spanish cultural influence was important in England, there was still much that was positive in the English image in Spain. One could not, of course, expect the English language or English books yet to have achieved any currency in Spain: England could hold only a pretty dim carbuncle of cultural achievement to the gleam of Spain's Golden Age, but it was possible for Spaniards to evince considerable respect for their enemies. Spain's most inventive poet, Luis de Góngora, canon of Cordova cathedral at the time of the Armada, alluded to England in his poem of praise for the departing fleet in terms reminiscent of the encomium of Shakespeare's Gaunt to 'This royal throne of kings'. He saw England as 'a once Catholic and powerful isle, . . . a temple of the faith, light of Mars, school of Minerva . . . happy mother and obedient slave of Arthurs, Edwards and Henrys, who were rich in fortitude and in faith'. True, this once glorious destiny had been perverted: the temple of faith was now a den of heresy; the once golden laurels were now fit to be replaced by a garland of dead grass; fame was turned to eternal infamy. Interestingly, however, the English escaped the blame, in Góngora's eyes, for what had befallen them. They were the victims of a wicked queen 'ruling you with a hand that grasps no sceptre nor sword but bone; many men's woman and sterile by all; infamous queen—not queen but rather she-wolf, lustful and beast-like! May fire of heaven descend upon your locks!' The effort to represent the war as directed against a tyrant rather than against the innocent people of the tyrant's land is familiar today. The same argument was used to express the nature of wars against Napoleon and Hitler; recently, Argentine propaganda represented the Falklands war as waged not against Britain but at the 'pirate, witch and murderess' at the nation's helm. Góngora was not alone in trying to justify the Armada in this way. Lope de Vega saw it as the struggle of a Christian Ulysses against a deceitful siren. Cardinal Allen's propaganda, though addressed to his fellow Englishmen, reflected a Spanish perception. Tainted by bastardy, Elizabeth was dyed deep by lust. She had 'abused her body, against God's laws, to the disgrace of princely majesty, and the whole nation's reproach, by unspeakable and incredible variety of lusts'. The legitimacy of her rule was thus doubly impugned, and by self-pollution she had turned her reign to tyranny. Elizabeth's image-makers had done their own work only too well, and all the enemy's rage was focused on the icon they had created.[14]

It must be doubted, moreover, whether élite perceptions of the religious nature of the conflict were shared at a lower social and educational level. Ordinary participants in the events of the Armada were either indifferent to the nature of the enemy or, in most of the instances we know about, saw him in profane terms as a monstrous caricature. If religious criteria operated at all, it was to define the enemy as 'irreligious'—or 'infidel' in the case of the sea shanties—rather than heretical; the objections raised were to his general barbarism, not his doctrinal errors. Each side had a 'Black Legend' of the other. According to English rumour, the Spaniards were coming in ships laden with instruments of torture, two kinds of whips, one for men and the other for women, and 'a great store of halters to hang all Englishmen'. 'The strange and most cruel whips, which the Spaniards had prepared to whip and torment English men and women' inspired an English ballad. Perhaps this sort of thing was believable to men whose knowledge of the effects of Spanish rule in the New World was derived from the lurid woodcuts of Theodore de Bry, depicting the conquest as a contest in mutual torture, or the invective of Father Las Casas's *Brief Relation of the Destruction of the Indians*—a work of impassioned homiletics, translated into English in 1583, in which horror stories exclude history. The example of the New World was evidently a present one to English minds. Burghley's pamphlet addressed, ostensibly from an English Catholic, to Bernardino de Mendoza, argued that the forcible conversion of England would be contrary to canon law—echoing one of the great themes of the Spanish debate about the legitimacy of conquest in America.

England seems to have been gripped by something very like a *grande*

A Declaration of the Sentence and depofition of Elizabeth, the vfurper and pretenfed Quene of Englande.

S IXTVS the fifte, by Gods prouidence the vniuerfal paftor of Chriftes flocke, to vvhome by perpetual and lavvful fucceffion, apperteyneth the care and gouernemēt of the Catholike Churche, feinge the pittyfull calametyes vvhich herefy hath brought into the renoumed cuntryes of Englande and Irelande, of olde fo famoufe for vertue, Religion, & Chriftian obedience; And hovv at this prefent, through the impietie and peruerfe gouernemēt of *Elizabeth* the pretenfed Quene, with a fewe her adhcarētes, thofe kingdomes be brought not onely to a difordered and perilloufe ftate in them felues, but are become as infected members, contagious and trublefome to the whole body of Chriftendome; And not hauinge in thofe parts the ordinary meanes, vvhich by the afsiftace of Chriftian Princes he hath in other prouinces, to remedy diforders, and kepe in obedience and ecclefiaftical difcipline the people, for that *Henry the* 8. late kinge of Englande, did of late yeares, by rebellion and reuolte from the See Apoftolike, violently feperate him felfe and his fubiects from the cōmunion and focietie of the chriftian comon vvelth; And *Elizabeth* the prefent vfurper, doth continevve the fame, vvith perturbation and perill of the cuntryes aboute her, fhevvinge her felfe obftinate and incorrigible in fuch forte, that vvithout her depriuation and depofityō there is no hope to reforme thofe ftates, nor kepe Chriftendome in perfect peace & trāquillety: Therfore our Holy Father, defyringe as his duty is, to prouide prefent & effectuall remedy, infpired by God for the vniuerfall benefite of his Churche, moued by the particuler affection vvhich him felfe and many his predeceffors haue had to thofe natyons, And folicited by the Zelous and importunate inftance of fundry the moft principall perfones of the fame, hath dealt earneftly vvith diuers Princes, and fpecially vvith the mighty and potent *Kinge Catholike of Spaine*, for the reuerence vvhich he beareth to the See Apoftolike, for the olde Amity betvvene his houfe and the Croune of England, for the fpecyall loue which he hath fhewed to the Catholikes of thofe places, for the obteyninge of peace and quietneffe in his cuntryes adioyninge, for the augmentinge and increafe of the Catholike faith, and finally for the vniuerfall benefite of all Europe; that he vvill employe thofe forces vvhich almighty God hath giuen him, to the depofition of this woman, and correctiō of her complices, fo wicked and noyfome to the worlde; and to the reformation and pacifrcation of thefe kingdomes, vvhence fo greate good, and fo manifold publike commodeties, are like to enfue.

AND to notefy to the vvorld the iuftice of this acte, and giue full fatisfaction to the fubiects of thofe kingdomes and others vvhofoeuer, and finally to manyfeft Gods iudgements vpon firne; his Holynes hath thought good, together vvith the declaratory fentence of this vvomans chafticement, to publifh alfo the caufes, vvhich haue moued him to procede againft her in this forte. F I R S T for that fhe is an Heretike, and Schifmatike, excōmunicated by two his Holines predeceffors; obftinate in difobedience to God and the See Apoftolike; prefuminge to take vpon her, contrary to nature, reafon, and all lavves both of God and man, fupreme iurifdiction and fpirituall auctority ouer mens foules. S E C O N D L Y for that fhe is a Baftard, conceyued and borne by inceftuous adultery, and therfore vncapable of the Kingdome, afvvell by the feuerall fentences of *Clement the* 7. and *Paule the* 3. of bleffed memory, as by the publike declaration of Kinge Henry him felfe. T H I R D L Y for vfurpinge the Croune vvithout right, hauinge the impediments mentioned, and contrary to the auncyent acorde made betvvene the See Apoftolike and the realme of England, vpon reconciliation of the fame after the death of S. *Thomas of Canterbury*, in the time of *Henry the fecond*, that none might be lavvfull kinge or Quene therof, vvithout the approbation and confent of the fupreme Bifhopp: vvhich aftervvard vvas renevved by kinge

This broadsheet declaring the deposition of the queen was intended to be posted in public places all over England had the Armada succeeded. Though there is no external evidence of authorship, the text seems marked by Cardinal Allen's hand, broadly following the lines of Pope Pius V's Bull of Excommunication of Elizabeth, but emphasizing the bastardy of a usurper 'conceyved and borne by incestuous adultery'. The final paragraph, ensuring collaborators of a plenary indulgence, seeks to convey the impression that the Armada was an official crusade, sanctified by the specific grant of papal indulgence to participants. In fact, however, Sixtus V was careful to confer no such special privileges on this particular expedition; the only indulgences available accrued from earlier bulls. In a similar attempt to enhance the Armada's impression of sanctity, a list of indulgences and pardons was printed at Lisbon on 6 April 1588 and circulated to recruits. None, however, specifically related to the Armada (see p. 38).

peur, induced by the long wait for invasion, while beacons were watched and bands mustered, as well as by the foe's reputation. Anxiety about vagrants is a sure sign of this syndrome. With the Armada in the Channel, county authorities were circulated with uncompromising orders from London, that 'no other persons be suffered to assemble together besides the ordinary bands'. Deputy lieutenants were to 'stay and apprehend all vagabond rogues and other suspected persons that are like to pass up and down to move disorders; and if any such be found with any manifest offence tending to stir up trouble and rebellion, to cause such to be executed by martial law'. Apparently, the authorities feared a recusant plot in the guise of a beggars' rebellion, a Bagaudae revolt.[15]

If, on the English side, fear took the place of religious fervour in inspiring resistance, the motive force on the Spanish side may have been hope of

plunder. Medina Sidonia's draft exhortation to his men, already quoted for
its passages of religious transport, appeals unashamedly to this unspiritual
inducement. After running through the great reasons of state which have
combined to launch the fleet—'glorious to our country, because God has
deigned to make it His instrument for such great ends; necessary for the
prestige of the king and the preservation of the bullion fleets; profitable
because the war in Flanders will be ended and we shall be saved the drain
of blood and substance which it draws from Spain'—he addressed directly
the self-interest of the men, adding 'profitable also because of the plunder
and endless riches we shall gather in England and with which, by the
favour of God, we shall return, gloriously and victoriously, to our homes'.
An Italian memorandum of July 1587, captured by the English and
endorsed in Burghley's hand with the words, 'Discourse to move the King
of Spain to enterprise some force against England', predicted that soldiers
would readily volunteer for the sake of the rich booty in a kingdom reputed
to be the richest and fairest (*il piú deleitoso*) in the world. An English
agent's report, filed a few days before the Armada sailed, claimed that 'the
Englishmen that be in Spain do report very foul speeches of the Queen's
Majesty; and they and the Spaniards desire but to set foot on land and all
shall be theirs ... The soldiers and gentlemen that come upon this voyage
are very richly appointed, assuring themselves of good success; in so much
as they might take up any wares there to repay it upon the booty they
should take in England.' When the fighting was over, Luis de Miranda
reflected, with his ship off the north coast of Scotland, that 'we all thought
we would emerge rich from this campaign and after what has happened
we shall leave it with only the shirts on our backs'. The letters of a Spanish
soldier, captured by the English, show the same aspirations. 'Do you
commend me in your prayers to God,' he tells his correspondent, 'that he
give me a good voyage and fortune that I may get the wherewithal to
repair my house and live at my ease.' His only fear was that the English
would surrender quickly and keep the spoils he craved. Again he wrote,
'Do you pray to God that in England He doth give me a house of some
very rich merchant where I may place my ensign, which the owner
thereof do ransom of me in thirty thousand ducats.' This was the standard
dream of the *conquistador*, the motive that had spread Spanish victories
from Naples to the Philippines, from Granada to the New World. A
repartimiento, a share not just of plunder in the ordinary sense, but of the
great fixed resources of immovable wealth which fell, by conquest-right,
to the victors, would be the soldier's indemnity and reward. Ordinary
booty could burst one's purse: in a single day's action in Sicily, Alonso de
Contreras once earned a hundred times a day's pay; but within a few days

he had squandered the lot. The aim of a *repartimiento* was quite different: it could permanently transform the *conquistador*'s way of life, changing him into a gentleman and enabling him to 'live at his ease'. Thus could the social ambitions which, as we have seen, were part of the common mentality of Spanish soldiers, be fulfilled. The fear the Spaniard felt was nothing like the *grande peur* of England. 'But I do fear,' his letter went on, 'that [the English] seeing us will presently yield and agree unto all that the king will demand of them, for that the king's force is marvellous great as well by sea as by land.' Effectively, it was not so much Protestantism and Catholicism that divided the forces that clashed in the Armada campaign as the secular sentiments of respective fear and hope.[16]

Three

The Shipboard Life

'THINK of it, Major,' says the hero of one of George Birmingham's novels, 'great fat dubloons ... And very likely there'll be gems, golden goblets with precious stones stuck in them. Those Spaniards were awful dogs for luxury.' Most Armada treasure-hunters have shared illusions of this sort. Spanish society—more, indeed, than that of England at the time—was ruled by austere aesthetics, puritanical values, and sumptuary laws. Scavengers managed to purloin 'certain chests of treasure and other things of good value' from the *Nuestra Señora del Rosario* of Don Pedro de Valdés, when she was taken as a prize and brought into Weymouth; but she was a pay ship, carrying 50,000 ducats of the king of Spain's, and must have made richer pickings than most. Irish beachcombers beat their English enemies to the scuppered embers of the *Rata Encoronada*, grounded near Ballycroy in September 1588 and—so the English believed—'took out of the wreck a boat full of treasure, cloth of gold, velvet etc.'. Spanish prisoners stripped in England were debagged, in particular instances, of breeches of satin, velvet, and, in two cases, cloth or lace of gold. But most of their plunder was too mean to be worth enumerating. Among modern treasure-hunters, even that most circumspect of marine archaeologists, Robert Sténuit, expected to find gorgeous tableware and was disappointed to prise only a copper saucepan-handle from the black amalgam at the bottom of Spaniards' Bay, where he had hoped for a dinner plate of gold. The wreck of the *Santa María de la Rosa* did yield a dinner-service, the property of the infantry commander Captain Matute; but, though fine in its way—it was 'personalized' with the owner's name-stamp—it was of mere pewter. A few silver-gilt vessels have been found, though most were probably intended for liturgical use.

Nevertheless, it is clear that the gentlemen-adventurers who sailed with the fleet carried a modest amount of wealth with them in the form of coin, gold chains—the most secure method of conveying personal wealth—personal jewellery, and silver tableware. The evidence comes from the wreck of the *Girona*, which sank off Port na Spaniagh, County Antrim, on 26 October 1588, in a desperate effort to reach Scotland before foundering. By the time of the disaster, the *Girona* had gathered up the crews

Captain Matute's name, engraved on the rim of an item from his pewter dinner service recovered from the wreck of the *Trinidad Valencera*. Most recovered tableware is of pewter, though the *Trinidad* wreck has also yielded a silver fork and a Ming bowl, and fragments of silver forks and tankards were yielded by the wreck of the *Girona*.

of four other abandoned ships, including the ship's company of Don Alonso de Leiva, who had the biggest complement of gentlemen-adventurers in the Armada—'all the noblemen in the expedition', according to Bernardino de Mendoza.

The *Girona* yielded her treasures to the excavations of Robert Sténuit in 1968. The exceptional quality and value of the artefacts he found reflects the exceptional wealth and social status of the victims. Some evoke the physical mores of the time, like the golden tooth-and-ear-pick in the shape of a dolphin; of others the significance is sentimental, like the charming gold rings inscribed with his grandmother's name and the legend, 'I have no more to give thee', worn at his death by Jean-Thomas Perrenot, nephew of Cardinal Granvelle. Some are of purely intrinsic value: the divers brought up eight gold chains, of which only one was ornamental, while of the rest, mere all-in-one money belts, the longest was over eight feet long and weighed more than four pounds: Francisco de Cuellar described gentlemen flinging their gold chains and coins into the sea at the wreck of his ship, and two others, at Streedagh in September 1588. Other pieces of goldsmith's work recovered by Sténuit were badges of rank, like the medallion of a knight of Malta, worn by the galleass's own captain, Fabricio de Spinola, or the gold dolphin-shaped whistle, which must also have been the property of a professional naval captain rather than of one of the gentlemen-adventurers. Yet others were devotional or talismanic: most of the soldiers or sailors carried a religious medallion, reliquary, or charm, and the finest of these were of gold or, like many from the wreck of the *Girona*, of cameo or intaglio-work set in richly chased gold. The most lavish of them all is of quite distinct type, in the form of a tiny gospel-book binding, chased with scrolls and flowers and suspended on a chain; St John the Baptist is depicted on the cover and a space for a relic is hollowed out on the reverse, while inside five tiny round pockets are moulded to take offerings of Easter candle-wax.

Of greater value in crude terms were the tokens of the pagan Renaissance that were carried alongside the devotional bric-à-brac. Personal vanity was out of fashion in Spain and most of the high-quality secular jewel-work—some three dozen examples retrieved from aboard the *Girona*—was probably intended as movable wealth in a highly concentrated form, carried by men who were also used to a good deal of spare cash in their pockets: Alonso de Luzón claimed to have lost 3,000 ducats' worth in all—ten times what would be thought a knightly income. The divers have raised 405 gold coins and 756 of silver from the wreck of the *Girona*. The most spectacular find, however, fits no obvious category: the wonderful set of portrait medallions of Byzantine emperors carved in lapis lazuli and set in finely worked gold, decorated with the faces of voluptuaries and curlicues of strapwork, studded with perfect pearls. These twelve Caesars of the *Girona* were too rich and too exquisite to take on a long and hazardous journey without good reason, too costly as a set to break up for cash. They must have been intended as a gift to some great potential English quisling, a noble collaborator who might be expected to welcome the Spaniards or at least to work with them. And probably, since the gift bore little trace of any religious connotation, the prospective recipient would not have been a Catholic.[1]

The most coveted of shipboard luxuries was a bed of one's own; all but the most exalted members of the ship's company had to doss down on the deck as best they could; there were no fixed sleeping quarters and hammocks were not favoured. In summer, the season of the Armada voyage, the favoured positions were on the upper decks, which were also the gun decks, for it was Spanish practice to mount the guns as high as possible because of the smoke emitted from the black powder they used. Although it was therefore important to keep these areas clear for action, canny voyagers liked to fit truckle-beds to the deck where they could and build little partitions to give themselves some privacy and to stake out a claim to deck space. In the interests of efficiency in battle, this was absolutely

This reliquary in the shape of a book, made of finely chased gold, contained an Agnus Dei, votive offerings of Easter candle-wax from St Peter's mixed with chrism and blessed by the pope. It was possibly a papal gift to the Bishop of Killaloe, who was aboard the *Girona* when she foundered.

From the *Girona* wreck, a pendant with the cross of the Order of Alcántara and St John of the Pear Tree, one of the Order's celestial patrons, originally set in green enamel.

prohibited on the Armada ships. Infringements of the prohibition began long before the fleet set sail: 'Before the death of the Marquess of Santa Cruz,' the king complained, 'I was told that because a large number of gentlemen and private adventurers were sailing with the Ármada, each of them presumed to erect chambers and lodging-spaces for themselves aboard the ships, and this was a great inconvenience because they took up space needed for action and impeded the mariners who had to sail the ships.' It may be that under the command of Medina Sidonia the practice was curtailed for a while, but the discipline had lapsed by the time the Armada—or what was left of it—got home. The decks of the returned ·ships were a maze of small enclosures fenced off for sleeping. Because the loftiest spaces were the most coveted, competition could be keen and sometimes violent. The mariners of Guipúzcoa claimed to have a customary right to the most sheltered places in the castles of their ships, from which the soldiers, they complained, excluded them 'by force': there was no justice in the allegation. Aboard the Armada, the castles were assigned as quarters to the soldiery for the very practical reason that it was from there that they would have to fight.[2]

So those 'awful dogs for luxury' carried and enjoyed the tokens, rather than the amenities, of civilized life. Their riches were costly, but not comfortable, their jewels fine but few. Medina Sidonia declared that the 8,000 freshly confessed and communicated men who took ship with him from Lisbon were 'the greatest jewels I carry': this was a representative expression of the genuinely austere atmosphere aboard and the genuinely spiritual priorities with which the Armada sailed. The shipboard day was rigorous and its consolations mainly religious. The passage of time was marked not by the modern practice of sounding bells but by the lilting of boys. It was one boy's work to mind the sand clock by which time was measured in 'glasses', each of a half-hour's length. Including spares, every ship would carry several of these, for the careful measurement of time was essential to accurate navigation. Only by making a practised estimate of the ship's speed, and comparing it with the time elapsed, could a navigator tell how far he had travelled on the open sea. Each turn of the clock was announced by a traditional lilt, probably similar or identical to those recorded by Eugenio de Sálazar on a voyage to Santo Domingo in 1574. After the first half-hour of every watch, for instance, the cry rang out,

> One glass has gone.
> Another's a-filling.
> More sand shall run,
> If God is willing.

To my God let us pray
To make safe our way,
And His Mother, Our Lady, who prays for us all
To save us from tempest and threatening squall.

During the night the boy would exchange calls with the members of the watch to make sure all were alert. At the seventh turn of the glass, the boy in charge sang out a warning to the next watch to be ready to change at the next turn; after eight glasses, or four hours, as in modern practice, he called, *Al cuarto, al cuarto, señores marineros!* 'On deck, on deck, gentlemen mariners of goodwill, on deck, quickly on deck, all those of Mr pilot's watch, for it's time now. Wake up! Wake up! Wake up!' Important members of the watch, whose vigilance was vital to the safety of the ship, including the boy at the glass, would rotate at intervals during the four hours' turn of duty. There were 'good morning' and 'good night' lilts, the first at daybreak, the second at the lighting of the ship's lantern, each accompanied by an appropriate prayer.

Blessed be the light so good,
And blessed be the Holy Rood,
And blest the Lord of Verity,
And the Holy Trinity,
Blessed be our souls—God save them—
And Blessed be the God Who gave them.
Blessed be the light of day,
And God Who with it lights our way,

greeted the dawn; night was heralded with a more general hope for a safe voyage, with a reassuring word of report: 'The watch is set, / The glass runs yet, / Safe on the seas, / If God decrees.'[3]

The changes of watch, the lilting of the boys, made shipboard life reassuringly conformable to predictable rhythms. The other great fixed points were the daily common meal and the services of prayer. These were particularly important if disciplined routine and high morale were to be maintained in an expedition like that of 1588, when there were huge numbers of landlubbers aboard, who took no part in the frequent secular rituals of the seamen. There had been amphibious operations on a comparable scale before—the ill-fated Tunis expedition of 1541, for instance, and Dom Sebastião's disastrous invasion of Morocco in 1579—but it had never before been necessary to transport soldiers so far in such large numbers. Many—perhaps most—must have found the experience frightening and odious. Fixed routines could impart some comfort in strange and dangerous surroundings. Thus it was no mere drill-book mentality that made Medina Sidonia order frequent drill and weapon training for the troops; nor was it religious sentiment alone that made him emphasize so strongly the need for regular religious observance.

The daily services were based on mariners' practice. Mass could not be said on board ship. The Spanish regulations repeatedly forbade it, on the grounds that the rolling or pitching of the ship might cause the Blessed Sacrament to fall or even to be pitched overboard. In the army, by contrast, daily mass was *de rigueur*—centuries before it became common practice among the Catholic laity. The pattern of religious observance in the fleet was therefore a new experience for many of the first-time shipmates. A last communion before embarkation had to be offered to the men at a great open-air mass on an island in the Tagus estuary. Once on board ship, the soldiers would find that the singing of the office was led, in accordance with maritime tradition, by the ships' boys. Despite the large number of religious aboard the Armada—all but the smallest ships had at least one priest aboard—the leading role of the boys' voices is specifically confirmed by Medina Sidonia's directions: 'Every morning at daybreak the boys, according to custom, shall sing their Salve at the foot of the mainmast, and at sunset the Ave Maria. On some days, and every Saturday at least, they shall sing the Salve with the Litany'—that is, the Litany of Our Lady. The Spanish Atlantic fleet held firmly to the doctrine that God preferred praise out of the mouths of babes and sucklings, and raised the voice of innocence aloft. The Marian emphasis of the services—especially on ordinary days, when the Credo seems usually to have been omitted— was also a naval tradition. Stella Maris was, for sailors, the most heartfelt of Our Lady's advocations, and Marian devotion was a meeting-place of the traditional, propitiatory religion of seamen with one of the great universal cults of the church.

The morning service followed the boys' lilt of daybreak and consisted of the Pater Noster and Salve Regina. On Eugenio de Sálazar's ship, these were sung by a boy without congregational participation, and this may have been the practice of the Armada, too. A lilt to bid good morrow followed, before the company dispersed to the duties of the day. The evening office was more elaborate. On ordinary days, the ritual began when a boy brought the newly lighted lantern on deck, singing his evening lilt: 'Amen and God give us good night. May the ship make a good passage and have a safe voyage, Captain, Sir, Master, and all the company.' The office, which had to include the Ave Maria, and which probably also featured the Pater Noster and Salve Regina, was then recited by two gromets.

Eugenio de Sálazar has left a good account of the full service for the eve of a feast-day or of a Sunday. An altar was set with candles and images—perhaps a triptych or travelling altarpiece—and the ship's master called in a loud voice, 'Are we all present?', to which the men responded, 'God be with us.' The master then chanted, 'A Salve let us say, to speed us on our way. A Salve let us sing: a good voyage may it bring.' After the Salve and Litany of Our Lady, the master might introduce the Credo by saying, 'Let us all profess our Creed to the honour and worship of the Blessed Apostles, that they may pray for us to Our Lord Jesus Christ to grant us a good voyage.' The Ave Maria would then be invoked by a boy chanting, 'Let us say our Ave, "Hail", For the ship and all who in her sail,' with the other lads responding, 'Blessed may she be.' The service ended with the usual evening lilt. For the Salve and the Litany, all voices would unite in one thunderous cacophony, 'a tempest of hurricanes of music' and 'a babble of braying', as Sálazar said of the croaking of the sailors aboard his ship in 1574, when only a single ship's company was to be heard. The noise raised aboard the Armada at Evening Prayer must have been awesome.

These were the great concerted devotions of the army and fleet. The individual piety of members of the expedition can be glimpsed in their writings, where they have left them, and in the profusion of amulets and other religious tokens which they carried on board. The marine archaeologists have raised touching examples of cheaply mass-produced medallions of the Virgin and saints, usually cast in pewter, but with many of copper and lead, which doomed voyagers wore round their necks and carried with them to the bottom. The religion of seamen has always veered towards superstition: that is the natural consequence of constant exposure to an implacable element. Ex-voto models of ships are one of the most important sources of information on ship design in the period, thank-offerings for deliverance from shipwreck, like the clothes Horace hung in Neptune's temple. None of the accounts of Armada shipwrecks or near-shipwrecks to have survived specifically mentions vows in peril, but all record invocations of the Virgin Mary, usually combined with evidence of a practical eye to the material chances of survival. According to the diarist of the *Gran Grifón,* the crew called 'on the Virgin Mary to be our intercessor in that most bitter journey, not neglecting withal our labours at the pumps and with the buckets'.[4]

It seems possible that soldiers, looked at collectively—though many wore medallions as charms—may have had a better grasp of Christian verities than most ordinary seamen. In part, this may have been because, as we shall see, they enjoyed the ministrations of a large and active corps of chaplains, but also, perhaps, because there may have been a sense in

Ordinary soldiers and seamen aboard the Armada carried modest talismans—pewter and copper medallions of sacred subjects, easily effaced by the seas in which their owners drowned. This example from the wreck of the *Trinidad Valencera* bears an image of Christ and the barely legible inscription 'Salvator Mundi'.

which soldiering attracted, from Spain, specifically 'religious' types. It is commonly said—and may be no less true for that—that the crusading tradition lasted longer in Spain than in the rest of western Christendom because of the protracted presence of confessional foes on Spain's soil and shores. The element in the crusading tradition which distinguished it from other ideological justifications of war was that it represented fighting in a holy or at least a formally hallowed cause to be an acceptable way of salvation for a layman—almost an ersatz work of mercy. I know of no texts to show that this particular way of looking at the soldier's profession was still alive in late sixteenth-century Spain. But, however that may be, it seems clear that arms were attractive to men of frustrated religious vocation or of strong inclinations towards religiosity. Jerónimo de Pasamonte, for instance, claims explicitly to have been led into soldiering by his want of means to train for the priesthood. Nor did he lose in the camp his longing for the altar. Captured by the Turks he tired his fellow-prisoners with his extemporized sermons and wearisome recitations from the divine office. When his soldiering days were over, he continued to belabour his superiors and his readers with many pages of sententious advice and popular theology, such as might have come out of any rustic pulpit. His fellow autobiographer, Alonso de Contreras, in one of his most extraordinary passages, tells us of the seven months he spent as a hermit in an interlude of his soldier's life, claiming never to have been so happy as in his poor and lonely cave—and this from a notoriously libidinous and hard-living picaroon who seems to have been addicted to his own adrenalin. Contreras's religious calling may have been shallow, or even insincere, but it was not unbelievable: it was representative of a common dilemma and of a surprising affinity between the rival claims of spiritual and worldly warfare.[5]

Seamen from Spain were used to sailing without benefit of clergy. Theirs was a world in which, as Eugenio de Sálazar said, the ship's master was a 'priestling' and the gromet a 'little monk'. When threatened with sudden death in a storm, they heard each others' confessions—the second Lazarillo de Tormes at his fatal shipwreck was unable to reciprocate because his mouth was stuffed full of food in a desperate effort to enjoy a last good meal. The exception to this dearth of clergy was the royal galley service, which played a negligible part in the Armada campaign. Galleys had chaplains aboard ship from the 1520s onwards. Regulations of 1557 showed special concern for the spiritual welfare of the oarsmen, who were obliged to confess in Lent and to be allowed to hear mass. It may be, therefore, that galley chaplains were like prison chaplains, with a special ministry to the cutthroats and desperadoes, captives and criminals who manned the oars, supplying a religious opiate to potentially mutinous men and salving the consciences of those in authority. In any case, the galley service was a world apart from the ocean-going navy. The galleys' main role was in short-range and usually coast-bound warfare. The problems of provisioning the vast crews compelled them to put frequently into port; and on active service they carried large numbers of soldiers on board. In short, galleys were a virtually 'amphibious' service, operating between land and open sea, with large numbers of landlubbers—both soldiers and oarsmen—swamping the professional mariners. The ocean galleons which formed the spearhead of the Armada had none of the limitations or encumbrances of the galleys, and belonged to a tradition that knew no need of shipboard priests.

The fact that the navy had no tradition of chaplains on board, that the office was chanted by the boys, and that the army's ranks were full of clerics *manqués* whose vocations seem somehow to have gone awry, may suggest that there was little work for the Armada's resident clergy to do. The Spanish army, however, required a vast organization of chaplains to minister to its spiritual needs. And an expedition like the Armada, which was a great missionary enterprise and virtually an act of divine propitiation, as well as an invasion force heading for an almost priestless land, had to take a large clerical contingent with it. The most famous and frequently reproduced image of the Armada—the presumed 'tapestry cartoon' of the National Maritime Museum, Greenwich—makes this point vividly. The most prominent individual figure in the composition, near the centre of the foreground, standing out dramatically from among his fellow travellers, is a friar in a billowing habit, his arms thrown out in exhortation or despair. To some extent, this reflects the English perception of the Armada as an engine of Catholic vengeance, launched to subvert the English Reformation. But that was, after all, the genuine and avowed nature and purpose of the 'Catholic Fleet', and the prominence of the friars is demonstrated, equally but more prosaically, by written sources too. Most of the official tallies assign 180 friars to the 130 ships of the fleet; Juan de Vitoria, who is usually cavalier about figures, but was reliant on clerical informants, seems unusually rigorous in contriving a total of 191.[6]

A glance at the Army of Flanders suggests that at the time of the Armada army chaplains were establishing an important place and a vital relationship with their congregations in the Spanish world. This was an important instance of a much bigger phenomenon. The growth of Catholic evangelism in the sixteenth century not only reclaimed lands lost to Protestantism but also took missionaries to the remotest corners of the world, as well as into under-evangelized parts of Europe: the neglected levels of society, the rural byways, the new urban slums. An army like Spain's, with relatively large numbers of long-serving professional soldiery, gathered together in forms of organization that made the men accessible to the ministry of the clergy, was an ideal field in which to sow the word. The Army of Flanders had the additional benefit of a commander like Parma, genuinely anxious for the spiritual welfare of his men, and a chaplain-general of exceptional gifts, Francesco de Umara, who shared the spirit of the times and sensed the possibilities of an unprecedentedly active ministry. Nominally, above Umara, there was a papal legate to the army: the very existence of such an office, newly created at Parma's request, is eloquent testimony to the status which the work commanded. For the effective management of the corps of chaplains, Umara was Parma's personal choice. The chaplains' work grew, even as it became better defined. They were the ministers of the sacraments among their flocks; from their role as confessors grew another as advisers, and they commonly helped the men in drawing up, for example, even the purely secular dispositions of their wills. No doubt in partial consequence, army wills—a good series survives from the next century—seem to grow in intensity of religious feeling and perhaps in extent of devotional bequests. The chaplains often ran military hospitals; and Umara encouraged them to be preachers, too.[7]

Not every army chaplain exhaled a pure Counter-Reformation spirit. The Franciscan Fray Antonio de Granata, for instance, who had done six

This near-contemporary painting, often plausibly said to be a tapestry design, shows what is perhaps meant for the battle of 8 August. The composition is dominated by the great galleass in the foreground, suggestive of the tremendous impression galleasses made on some of the English who beheld them. The bank of oars is plainly visible, as is the heavy armament: the artist has managed to fit in representations of twenty-eight great guns: no other ship in profile is shown with more than twelve. Only the galleass's tail-guns, however, are engaged (see p. 168). A motley ship's company is shown on the deck, including a friar with his arms raised in exhortation or prayer, a jester, and a royal official tugging at his beard, as if to symbolise the fanaticism and folly which the English attributed to the enterprise. Note the papal banner flying from the topmast.

years in the job by 1588, was an unregenerate pre-Tridentine figure, who sang profane songs to the sound of his lute, conducted bogus and prurient visitations of convents, extorted gifts, wore furs and gold chains, and battered his denouncers. He claimed at his trial for these offences to be a martyr who spoke the truth and converted sinners. His persecution, he suggested, was 'to force me out, and my Franciscan brothers with me and replace us with Jesuits. But St Francis will punish the persecutors of his order.' He may have been voicing a genuine Franciscan anxiety. Certainly, the Jesuit mission in the army was growing more important, as the need for a more active evangelization of the men became increasingly felt among their commanders. The new, evangelically aware orders of the Counter-Reformation, of whom the Jesuits were the most conspicuous and dynamic, were attracted to the army for the same reason as they felt drawn to the slum-ringed boom towns like Seville and to the dense, servile native populations of the New World. Rootless masses were at once an easy and urgent target for their ministry. Commanders interested in the spiritual welfare—or, at least, concerned for the dogmatic instruction—of their men seem to have recognized the Jesuits' special gifts. In 1587, Parma's call for a central Jesuit mission to the Army of Flanders was answered by his personal confessor, Thomas Sailly. Within a generation, the Jesuits could claim to have enhanced the morals and galvanized the strength of the corps of chaplains as a whole.[8]

By comparison with those on the Spanish side, the sources about the religious life aboard English ships are exiguous. Spanish historians have tended to assume that the English were genuinely irreligious—rather, indeed, as contemporary Spanish propaganda depicted them. But arguments *ex silentio* are always dangerous, and, in this case, there is good reason to think that English practice was really very close to that of Spain. English commanders seem to have been as anxious as their Spanish counterparts to keep the fleet in a state of grace or—granting that this was impossible—to protect their men from some of the more common ravages of sin. As well as more practical purposes, shipboard discipline served the need to propitiate God. Medina Sidonia ordered his subordinate officers, that 'the desire to serve God ... should be kept constantly before the eyes of all of us, I enjoin you to see that before embarking, all our men be confessed and absolved, with due contrition for their sins. I trust that this will be the case with everybody, and that by this means, and through our zeal to serve God effectually, we may be guided as may seem best to Him in Whose cause we serve.' Consistently with these priorities he singled out swearing and blaspheming, before whoring and above shipboard brawls, and even above the gambling which, he claimed, caused many of the other excesses, as offences to be particularly checked. In these respects, English orders reflected much the same scale of values; only the punishments were more Draconian. An English serving-man, for instance, would get his pay docked for a casual oath or failure to attend divine service. A Spaniard would merely lose his wine ration for a mild blasphemy. Puritanism and 'the spirit of the Counter-Reformation' were, in many respects, kindred phenomena. And fighting men, whatever their provenance, were kindred spirits.[9]

No serious attempt to enforce the regulations against oaths and blas-

The pair of gaming cups recovered from the wreck of the *Trinidad Valencera* show that Medina Sidonia's ban on gambling was eluded. The cups could be fitted together and the dice rattled inside them.

phemies seems to have been made on either side: one might as well milk a billy-goat into a bottomless bucket, as a sixteenth-century phrase puts it. Dicing—a characteristic 'soldier's vice', as Juan de Vitoria called it— could hardly be suppressed, as the dicing cups which have survived from the wreck of the *Trinidad Valencera* show. Sexual offences were equally part of the way of life. Sailors may have been used to long periods of sexual deprivation; not so soldiers, and bordello-ships commonly followed troop fleets. Army regulations provided for four to eight prostitutes for every hundred men; perhaps ten to fifteen were tolerated in practice. In the Armada's case, because of the emphasis placed by the Spanish command on the need to engage divine sympathy, Medina Sidonia enjoined strict efforts to preserve purity. However, the prohibition of women on board did not, it seems, extend to married women living with their husbands, and according to some sources it was a case of sexual jealousy that caused the act of sabotage in which Martín de Oquendo's flagship was burnt. According to the informer, Francisco Valverde, 'it was said that some women had gone on board the Armada in the guise of men, whereupon the Duke of Medina caused a search to be made and found thirty'. Even if this detail is not strictly true, it shows how the commander's efforts to impose chastity on the fleet were received by the rank and file—at the level not of a pious aspiration but a dirty joke. The soubriquet given to one ship—'hulk of the women'—suggests again that the duke's instructions were evaded, derided, or both.[10]

The structure of worship, and the staffing of ships with divines, was much the same in the English as in the Spanish fleet. The English had the same daily and obligatory services at dawn and sunset, composed of psalms and prayers, as the Spaniards' were of prayers and canticles. Though there were fewer clergy with the English, those there were played a bigger part in the services, probably conducting them and frequently adding a sermon. We know of preachers aboard the *Ark Royal, Elizabeth Jonas, Bear, Rainbow, Lion,* and *Revenge.* The term 'preacher' implies a militant and radical form of Protestantism, which may have been very different from the religion of most Englishmen, but which must have contributed a measure of *odium theologicum* on those ships where it found a voice. The navy was already becoming the distinctly 'puritan' service it was to be in the first half of the next century.

If the English sources have relatively little to say about shipboard religion, that is not because devotion came lower on Englishmen's scale of values, but because of the different circumstances of the two fleets. The English were operating in home waters, with all the advantages for morale which that conferred. They did not have to accommodate large numbers of soldiers, out of their element and needful of reassurance. Nor did they need to be at sea for the long periods endured by the Spaniards. The secular rituals of life afloat were enough to sustain them, and the regular spiritual ministrations provided on board ship could meet their needs. Apart from the ministry of the preachers to the seamen, no missionary purpose was served by the English fleet and to carry divines in large numbers would have been supererogatory. Anxiety about morale, about relations between soldiers and sailors, and about discipline aboard, which all combined to emphasize the importance of formal, common devotions aboard the Spanish ships, were not present in the English case.

Apart from religious offices, the other great fixed ritual of the day was the common meal. How the cooking was done and service provided for such large numbers as the Armada ships carried remains a mystery. Generally, the cooking arrangements aboard sixteenth-century vessels defy attempts at credible reconstruction. The galley was an open fire laid of brushwood and charcoal on top of the ballast stones, down among the bilges in the most insanitary recesses of the ship, where rats and every kind of filth accumulated. A report by Sir William Wynter in 1578 found this location inconvenient, not because it was unhygienic but because it prevented access to timbers which were rotted by 'the leakage of beer with the shedding of water upon the said ballast', which, in English ships, was of gravel rather than the large stones and lead ingots favoured in Spain. Wynter's solution was to put the cook-rooms higher and use larger ballast-stones that would allow the air to circulate. Only a small fire was ever maintained because of the terrifying incendiary risks and it must be supposed that hot food was rarely if ever taken save by the sick and a further privileged few. The Armada ships had only a single huge cooking-pot each—the remains of that of the *Gran Grifón* were excavated from the sea-bed off Fair Isle in 1970. Only a small proportion of the ships' companies can have enjoyed the table service described by Eugenio de Sálazar, seated at a dingy cloth dotted with 'little piles of decayed biscuit' and resembling 'a field of stubble scattered with mounds of manure'. These fortunate few were greeted by the gromets' lilt,

Table! Table! Captain, Sir, and Master and good company! Table laid, dishes made! Water as usual for Captain, Sir, and Master and good company!

> Long live the King of Spain,
> By land and on the main!
> He who offers him war,
> May he have his head no more!
> Who does not say Amen,
> May he never drink again!
> To table with good speed!
> Who comes not, will not feed.

The lower officers, including the commander of the artillery—so often the despised service—sat with the common seamen and soldiers on the floor, paring bones 'as if they had studied anatomy in Guadalupe or Valencia', while the ship's notary and the captains commanding the companies of soldiery stood by, the notary at his portable desk, checking each other's accounts of the distribution of rations and entering them, signed and countersigned, in the ledger.[11]

The staple diet of both sides, English and Spanish, is well documented and makes a vivid contrast. The men of the Armada ate Mediterranean fare, of fish, oil, rice, and wine, for which the English substituted beef, butter, and beer. Bacon, cheese, hard tack, and pulses were the great common items. Each fleet, no doubt, had an appropriate diet: wine, we know, tended to make English soldiers sick; so, no doubt, would some of the exotic fish specified in the Spanish ration-lists, like squid and tunny. The Spaniards thought they were better off: northern climes and diet bred 'drooping hearts and lax and frozen bodies'. But closer examination seems to suggest that the English were much the better nourished of the rival forces, at least for the brief period while their rations lasted. In the first place, the Spanish rations were scanty, even by contemporary standards—at 4,000 calories per day they were barely more, in calorific value, than some Morisco bandits had received for their maintenance after capture in

1576, and somewhat less, man for man, than was provided on other major expeditions: almost fifty per cent less, for instance, than on Dom Sebastião's ill-fated African venture of 1579. In planning the expedition, the Marquess of Santa Cruz originally planned for a meat ration fifty per cent higher than that carried by the Armada. Moreover, the men rarely got as much as was envisaged in Medina Sidonia's orders: the actual wine ration seems to have been about half that envisaged—about two-thirds of a pint—and Medina Sidonia ordered it to be cut to even less in the case of Andalusian wine 'which makes those not used to it ill'. It should be added that unwholesome drink was the commonest reputed cause of sickness aboard ship and that the English had even worse trouble with their beer, which was poor from the start for want of hops and soured quickly. 'The mariners have a conceit,' wrote Howard to Walsingham on 26 August 1588,

and I think it true and so do all the captains here, that sour drink hath been a great cause of this infection amongst us; and, Sir, for my own part I know not which way to deal with the mariners to make them rest contented with sour beer, for nothing doth displease them more.

The purveyor's solution was to brew the sour beer over again with an admixture of new ale; Howard's, more sensibly, was to transfer the brewing to a new location where there were more hops. Water did as much damage as bad wine or sour beer—sometimes indirectly, as among Parma's troops who died in Flanders from fire-water or bad wine 'because they were unaccustomed to drinking water', but water was so hard to preserve in a palatable, even a sanitary, condition, that the common scruple against it was prudent. On the Spanish side, only the galley oarsmen, who got no wine ration, drank it unmixed save with the vinegar that was always added to prolong its life. The English ration of a gallon of beer a day rendered drinking water superfluous.[12]

In the early days of the campaign, English rations were probably not far short of what was intended. James Quarles's purveyancing estimates match Burghley's prescriptions item for item. For flesh days—Sundays, Tuesdays, and Thursdays—a pound of biscuit, pound of beef, and gallon of beer were to be provided; the Spaniards had no such protein-rich days. On Wednesdays, Fridays, and Saturdays, the same amount of biscuit and beer would be complemented by a quarter of a stockfish or an eighth of a ling, with a quarter of a pound of cheese and two ounces of butter. Mondays were reserved for a pound of bacon and a pint of peas—omitted from Quarles's account. By comparison, the diet on which the Spaniards started the campaign seems enfeebling. The daily staple was of one and a half pounds of biscuit, two pounds of bread when available, and a measure of wine. On Sundays and Thursdays, this was to be enhanced with six ounces of bacon and two ounces of rice; on Mondays and Wednesdays by six ounces of cheese and three of pulses; and on Fridays and Saturdays by six ounces of tunny, cod, or squid and five sardines, with oil and vinegar and three ounces of pulses. The tuna was supplied—at competitive prices—in part from Medina Sidonia's own fishery business, despite the havoc wreaked on his smacks and nets by Drake's cruise against the Iberian coast in April 1587. On paper, the English seem to have been much better provided. They were, after all, close to sources of victualling, and were able to turn over their supplies relatively quickly, with less reliance on highly durable products: salt beef, for instance, was more perishable than the cured cod and bacon favoured by the Spaniards (presumably, the *Rosario* was exceptional in actually having aboard more beef than fish). Even so, the Spanish preference for salt fish over salt meat may have been dictated by other considerations, for while Medina Sidonia thought fresh fish unhealthy, the Portuguese apothecary Henrique Días believed the same of salt meat. The English, finally, in addition to their other advantages, might have been thought to have better access to fresh substitutes for shipboard fare. Burghley could afford to be generous in planning and Quarles to be expansive in his accounting. In reality, however, the English were ill-organized by comparison with their adversaries and during the period of the fighting their nutritional advantage must rapidly have become negligible. By late July, the English were messing six men to a mess aboard ship: in other words, six men were sharing four men's rations; even at their best, the purveyors never seem to have got more than three weeks' victuals on board, and during the active part of the campaign the fleet lived from hand to mouth. By the end of the campaign, they were, as we shall see, hardly better off than the Spaniards who had provided for three times as many mouths, for more than twice the amount of time.[13]

When Medina Sidonia did cut rations, he was obliged to do so with a vengeance. It was hard to take spoiling into account when planning an expedition, because the hazards of seafaring, the season's delays, and the action of the enemy were unpredictable in their effects. In the case of the Armada the difficulties were compounded, of course, by the scale of the venture and by the particular short-term crisis in the availability of barrel-staves. The traditional view that this was Drake's work seems justified. He captured a vast quantity of coopers' wares off Sagres towards the end of May 1587: 'the hoops and pipe staves were above sixteen or seventeen hundred ton in weight,' he reported, 'which cannot be less than twenty-

The tunny fisheries of Cadiz, with the fish being netted, smoked for the marine trade, and stored. The Duke of Medina Sidonia had a large stake in the tuna industry and supplied the Armada from his own business. Some of the wine, too, came from his own family vineyards of Condado.

five or thirty thousand ton if it had been made in cask, ready for liquor, all which I commanded to be consumed into smoke and ashes by fire, which will be unto the king no small waste of his provisions.' It was at about this time that Medina Sidonia and Philip exchanged complaints of the want of pipes for wine and water.[14]

In trying to make allowance in advance for spoiling, Medina Sidonia reckoned to be 'on the safe side'. He took three times as much water as he expected to need. Even so, the provision proved woefully inadequate. On 11 August, he halved the water ration; the men were now getting only 240 grams of biscuit a day instead of the official ration of 800 grams. The fleet was fifty days out of Lisbon, when she had sailed with a nominal ninety days' victuals on board—without counting the opportunity for replenishment, and for relief of the stores by the use of fresh supplies in port, which had been gained from the delay in Corunna in June and July. Even at the new stinted rates of issue, there were only eight days' biscuit and wine left when Medina Sidonia reached home on 23 September.[15]

Some ships may have been better off: it must be remembered that the Armada was dispersed by storms on the way home and that after 16 September even the small nucleus that remained with the flagship was reduced to a policy of *sauve qui peut*. No ship was in a position to succour another, and the large fighting ships were suffering from the effects of the decision, prudently made at the start of the campaign, to lighten them by shifting their stocks to smaller vessels. The supply ships seem to have returned home in relative abundance: even in mid-October, for instance, an inspector could report that some of the returned bacon and salt cod was still edible—'considering,' he added, 'that it is old and that it has come all that way ... for it is not yet putrid, nor does it smell bad; only its appearance is decayed and in other respects it is good'. Unappetizing as it is, this report shows that basic rations survived on some ships for longer than had been anticipated. The same conclusion is suggested by a smug account in which the purser Pedro Coco Calderón commends his own good husbandship, dwelling with some pride on the surplus he maintained. One suspects his achievement may have been at the cost of the fighting ships. Some of the latter were much worse off even than the flagship. According to the examination of Emanuel Fremoso, for instance, by the time his own vessel, the fleet's vice-flagship *San Juan* of Juan Martínez de Recalde, came into the Blaskets off Dangean-ni-Cushey on 1 September, men were dying daily of hunger and thirst. It would be unwise to take this literally: Fremoso was in any case a disaffected Portuguese and was speaking under interrogation in the immediate, bitter aftermath of the débâcle; but it may be acknowledged that the shortages were severe and often fatal. Juan de Vitoria's horror stories include sailors who sold their clothes for a few drops of water—though he admits that some of the clothes were lost by reckless dicing. Towards the end of the voyage, if the complaints of Guipúzcoan sailors can be believed, the problems were exacerbated by a breakdown of discipline: soldiers stole the sailors' rations and used their superior weaponry to raid the stocks of victuals 'at will'.[16]

Even while full rations remained available, the diet was humdrum and unsatisfying. Eugenio de Sálazar's account of mealtime conversation aboard an Atlantic ship—composed entirely of wistful exchanges about the fresh fruits and vegetables one missed from home—becomes poignantly intelligible against the menu of dried and salted protein sources.

One man says, 'Oh, is there no one who might have a bunch of white grapes from Guadalajara?' Another, 'Is there anyone here who has a plateful of the cherries of Illescas?' Another, 'I could just do with a few turnips from Somo Sierra.' Another, 'A stick of chickory for me and an artichoke top from Medina del Campo.' And they all belch up their longings and disgruntlement over things they can't have in the place they're in. And if you want a drink in the midst of the sea, you can die of thirst, for they'll serve you your water by the ounce, like in an apothecary's shop, when you are full of dried beef and salt food. For my lady the sea will not suffer nor save meats nor fish that are not dressed in her salt. And everything else one eats is decayed and stinking like the blackamoor's boots.

One way of getting supplementary delicacies was to go sick. Spanish ships carried livestock to be slaughtered at need for the nourishment of the ill or wounded. The bones of sheep, chickens, and cattle found amid the embers of the wrecked galley of the *Trinidad Valencera* probably belonged to supplies of this sort, rather than representing the detritus of some lavish captain's table. The whole Brazil-nut—perhaps the most pathetic relic of

A Brazil-nut, 5.2 cm. long, recovered from the wreck of the *Santa María de la Rosa*, shows that exotic comestibles were carried aboard the Armada. The nut may have come from some gentleman's hoard of private victuals or from the stock of physic for sick and wounded men.

the Armada wrecks—from the *Santa María de la Rosa* is also likely to have formed part of the medical supplies, for nuts, raisins, and preserves were highly esteemed as physic. While the extraordinary fare lasted, fresh meat was served daily on the hospital ship—save for eggs or fish on fast days. All ships had live chickens in the galley for the feeding of the sick, for the chicken was the most prized of all meats for its recuperative properties. The fleet carried a range of other livestock—sources of beef and mutton, three calves and fifty sheep aboard *Nuestra Señora del Rosario*—to be slaughtered for the needs of the sick or wounded; and fruit conserves took the place of the fresh or dried fruit which was supplied almost daily to patients in hospitals at home. The meat ration was small—100 to 150 grams—but was intended as a supplement to regular rations. The English had no comparably well-organized system for the nurture of the sick, and had to obtain their extraordinary victuals by luck or skill. When the campaign was over, Howard wrote to Burghley explaining that he had taken the liberty of embarking 'certain extraordinary kinds of victuals, as wine, cider, sugar, oil and certain fish' at Plymouth 'to relieve such men withal as by reason of sickness or being hurt in fight, should not be able to digest the salt meats at sea, as also for the better lengthening of our ordinary victual'. He also admitted to extra beer and wine, which he would pay for himself. The goods, no doubt, were requisitioned from foreign merchantmen, and seem to suggest that the English thought the Spaniards' 'Mediterranean' diet—suggested by all the items except cider—

superior to their own. In May he had impounded another item which, to Spaniards, would seem regular provender: a cargo of rice from aboard the *Mary* of Hamburg, justifying his action on the grounds that he had only three weeks' supplies and 'I would to God I did know how the world went with our Commissioners'. We hear of no other exotic victuals on a large scale, though at the end of July Sir William Wynter was eating a present of venison bestowed by Walsingham—'the best store of victuals', he gratefully acknowledged, 'that I and Sir Henry Palmer have'.[17]

Faced with the inadequacies of distribution and the ravages of spoliation, Spanish victualling represented a creditable job of organization. The English, with less excuse, suffered, at an earlier stage, shortages worse than their adversaries, driven, according to an impartial source, 'to such extremity for lack of meat that my Lord Admiral was driven to eat beans, and some to drink their own water'. To belong to a bureaucratic system, however, such as kept the Armada fed, brought corresponding disadvantages, bewailed in the Guipúzcoan sailors' complaints of

the many hurts and griefs and vexations which the masters and owners and seafaring men of this province had received and daily did receive at the hands of the ministers, accountants, sutlers and other officials, who all in their offices had continually sought and did seek to wreck and ruin the said ships' masters and seamen of this province.

Basques then, as now, were quick to complain of the deficiencies of a centralized administration; but their accusations ring true with the resentment of independent men confronted by over-organization and inefficiency and of basically honest men faced with official peculation.[18]

Four

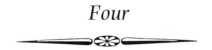

Strategy: Evolution and Confusion

LIKE most wars, the Armada campaign was fought for peace. And as with most wars, both sides foundered into it, bewildered about their own objectives. In most wars, however, initially rough-hewn strategies are gradually shaped into coherence. In the case of the Armada, that simply never happened. Both sides groped myopically in the confusion of muddled thinking throughout the campaign. Both suffered, without relief, from ill-defined, ill-informed, and internally contradictory strategies. Both were ruled by divided counsels from which a clear consensus never emerged.

There can be little doubt about both sides' remote grand strategies of peace. Peace was—after her own survival—the single most constant feature of Elizabeth I's policy. Her image as an eirenic queen was far more precious to her, personally, than her reputation as a Protestant prince. Spenser sensed this as he wrote the *Faerie Queen*:

> Thenceforth eternal union shall be made
> Between the nations different afore
> And sacred Peace shall lovingly persuade
> The warlike minds to learn her goodly lore,
> And civil arms to exercise no more.
> Then shall a royal virgin reign.

Soon after the Armada, with the war against Spain apparently at its height, the court play *Histrio-Matrix* made peace its major theme. Whereas, in the next century, servants of the Spanish monarchy would choose to be portrayed on rearing war-horses, Lord Burghley, in his most famous portrait, is mounted on an ass—the creature Christ chose to ride into Jerusalem to proclaim his reign of peace. From the moment of her accession, Elizabeth ruled determined to spare her subjects the cost of war and herself, therefore, the price of their resentment.

The provocations heaped upon Spain by English excesses in the years before the Armada were never intended to compromise the policy of peace. Elizabeth was, however, the victim of her own divided aims. She had an aggressive Protestant party at home, on whom her throne depended and whom she had to keep appeased; she acknowledged the wisdom of limiting

In contrast to Lord Howard of Effingham (p. 110) Burghley rides an eirenic beast. He holds a rose and a lily, symbols associated with the queen and appropriated, like much of Elizabeth's royal imagery, from the traditional iconography of the Virgin Mary. In Burghley's hand, they must be understood as tokens of his devotion to the queen. The same theme is suggested by the sealed knot and motto 'A single heart, a single road', inscribed on the tree-trunk: in Elizabeth's court avowals of service were typically expressed in quasi-amorous language. There is considerable emphasis on Burghley's Order of the Garter, which is shown twice in great detail.

Spanish strength; and she coveted the wealth of the Spanish monarchy. Hostility to Spain was therefore inevitable, competition irresistible. It was not easy, however, to tread on the edge of Spanish tolerance without overstepping its limits; nor, when Elizabeth had reluctantly embarked on a limited war, to keep it to the terrain of her own choosing. Her decision to meddle in the Netherlands from 1584, supplying troops for the aid of rebels against Philip II's rule, was made with the aim of enhancing her own security by keeping Spanish arms engaged. One effect of her intervention, however, was to convince Philip II of the argument, which he had frequently heard but consistently resisted, that there could be no peace in the Netherlands until England was conquered or cowed. Thus by trying to thrust her frontier against Spain to a safe distance from home, Elizabeth helped to bring war closer. Where a less resourceful or less determined enemy might have been distracted or deterred, Spain was provoked. English infringements of Spain's imperial monopolies in America and the Pacific were similarly ill-judged. In a sense, they were not directed against Spain, but were designed to secure for England a fair share of the profits of interloping in the uncontrollably vast Spanish Main, which otherwise would have been enjoyed without competition by French and, ultimately, Dutch pirates. In any case, English piracy on the Indies sea-routes and interloping in the New World trade were relatively small incursions into the king of Spain's domains. Yet it was impossible for Philip to tolerate them, not only because they compromised his precious 'prestige' but also because they were a constant drain of resources which it seemed cost-effective to try to end, once and for all, even at great short-term expense. Don Diego Pimentel told his Dutch interrogators that protection of the Spanish Main was one of the chief motives of the Armada because 'it were better (as he thought) to have a sure peace', by conquering or cowing the English,

> than that the king should continually keep such an army for to keep free the passage of the Indies, as he hath prepared now for to subdue and overcome the whole kingdom of England ... The occasions which made the king to undertake these wars were, that it was not convenient, that one Drake, with two or three rotten ships, should come always and at his pleasure, to spoil the havens of Spain and to rob the best towns thereof, and so to hinder the negotiation of the Indies.

The voice of a proud captive, stung into arrogance by misfortune, can be clearly heard in those lines.[1]

While Philip II did not share Elizabeth's obsession about peace, it would be wrong to suppose that he was a warmonger animated by *Weltpolitik* or avid to add to his dominions. If he was unable to give his subjects the long respite from war which those of Elizabeth enjoyed in the early part of her reign, it was not because he loved belligerency but because his vast dominions, with their long frontiers and innumerable hostile neighbours and rebels, were involved on too many fronts to preserve peace simultaneously on all of them. Philip had been king of England during his marriage to Mary I and the realm continued to enjoy his special favour and protection when Elizabeth came to the throne. Far from wishing to make war, Philip had enlisted all the arts of peace in his relations with Elizabeth, offering her marriage, seeking her conversion, invoking the long tradition of friendship between England and Castile and, above all, holding at bay the pressure for a Catholic crusade against her. By 1588, Philip was committed to overt hostilities. The negotiations he permitted in the Low Countries were a conscious charade, intended only to lull Elizabeth into relaxed vigilance. Yet he had embarked on a warlike course slowly and reluctantly, because English provocations left him no alternative; he adhered to it only because the expense of his preparations had to be justified by action; and it is likely that his objectives remained limited. In the secret instructions he issued to the Armada's commanders he specified three goals, short of the conquest of England, which represented the minimum acceptable return on his investment: English withdrawal from the Low Countries, peace in the Spanish Main, and toleration for Catholic worship in England. That was as much as he expected to achieve and, perhaps, no more or less than he seriously hoped.

When the Marquess of Santa Cruz first proposed to his master the formation of the Armada, he was in no doubt of its purpose: 'to make you king of that kingdom'. When at last it sailed, one of the most senior commanders aboard, Don Diego Pimentel, assumed that its goal was 'to subdue and to overcome the whole kingdom of England'. Yet a strong body of contemporary opinion doubted even whether Philip seriously proposed to invade, much less to conquer, England. The pope's mistrust of Spanish intentions was notorious. The Armada, he averred, had been raised only for 'a brag' and to frighten Elizabeth into peace. As reported by English intelligence in August 1587, he dismissed Philip as a deceiver who 'ever shut up in the Escorial, had no more care for the Catholic religion than a dog', and as 'a coward that suffered his nose to be held in the Low Countries by a woman, braved and spoiled at his own nose in Spain by a mariner'. The allusions, of course, were to Elizabeth and Drake. The pope, the report went on, then slipped into his characteristic ironic praise of Elizabeth, 'that there was no prince in the world showed any courage but one woman, for whom he would give all [the] treasure he had gathered together that she would become a Catholic'. All these remarks were perfectly credible, and entirely consistent with the pope's usual forthright language and coarse humour: he was even known to joke about the race of heroes he could engender if he married Elizabeth. He had given the Spanish ambassador the impression that he regarded the Armada as a traditional 'diversion abroad' intended to secure domestic peace—presumably in the Low Countries—by shipping out fractious noblemen; and, indeed, there may have been more than irony or speculation behind this remark, for Parma himself had once stressed this very purpose in a memorandum to the king. Sixtus promised a million gold ducats in aid to the Armada, to be paid as soon as the first troops landed, but only, as the Spanish ambassador perceived, 'in the belief (as I recollect writing to Your Majesty at the time) that the undertaking would never be carried through; and that it would serve him as an excuse for the collection and hoarding of money in all sorts of oppressive ways, particularly from subjects of Your Majesty'. The small trust, the ambassador warned, 'that can be placed in him may be judged by the little trust he places in us'.[2]

While the pope's testimony is suspect, other observers were equally sensitive to the hesitations of Spanish policy. The dispatches of the Venetian ambassador, Hieronimo Lippomani, reveal a mind struggling against disbelief in King Philip's seriousness of purpose. The king was 'naturally inclined to peace'; war was 'quite against his natural temper'. For a long time, Lippomani seems to have supposed that Philip was seeking peace by preparing for war, according to the adage; or that he was appeasing papal war-lust; or else trying to quell English provocations, such as interference in Flanders and the Indies and support for the pretender of Portugal, by a show of strength. 'What has enraged him more than all else,' Lippomani reported on one occasion in July 1586,

and has caused him to show a resentment such as he has never before displayed in all his life, is the account of the masquerades and comedies which the queen of England orders to be acted at his expense. His Majesty has received a summary of one of these which was recently performed, in which all sorts of evil is spoken of the pope, the Catholic religion and the king, who is accused of spending all his time in the Escorial with the monks of St Jerome, attending only to his buildings, and a hundred other insolencies.

At Elizabeth's court, entertainments were made to convey serious messages about royal policy; and Philip's reaction, if Lippomani is right, may have been inspired by more than personal pique. Yet this could hardly by any standards have been reckoned a convincing pretext for war. At times, Lippomani felt inclined to believe that the king really did intend to strike at England, for 'His Majesty cannot do less than punish the queen, if he desires to preserve his reputation and his possessions, for she is the foe of all good people, aye, and of God Himself'. Here, the ambassador acknowledged the force of two of the strongest points in the scale of values that underlay Spanish policy: prestige and Catholic self-righteousness. But by the time the Armada was actually being assembled, his doubts returned. 'Perhaps', he wrote of Philip, 'he hopes that the sailing of the fleet will greatly facilitate the conclusion of an honourable treaty.' By the time the expedition had actually set sail, he felt reluctantly obliged to acknowledge that Philip really meant to fight. 'Wise men wonder what can induce the king to insist that the Armada shall give battle to the English.' The conviction that God was on the Spanish side, he felt, had played some part, as had consideration of the cost of keeping the Armada in port. Above all, win or lose, the Armada seemed to present the shortest way to a definitive peace, for the English would be obliged to come to terms, even after a costly victory, whereas 'if they lose a battle they lose all at one blow'.[3]

The ambassador evidently felt baffled; yet the tenor of his dispatches was closer to the truth than he seems to have realized. His doubts closely reflected the indecisiveness of Spanish policy. The pope was wrong to think of the Armada as a mere gesture: as we shall see, it went equipped for invasion and, if need be, for conquest. But, while Philip provided for a range of eventualities, he settled for none. He was caught, like Zeno's ass, between competing objectives. And the Armada was prepared and dispatched in ignorance of its ends. For a long time, the king continued to hope that it would be unnecessary: the English would be intimidated into peace, and the Armada could then be used to reinforce the Army of Flanders. Or else the fleet's job would be done for it by an English or Scottish Catholic plot. Or else, balking at the major task, the Armada could be used against Ireland, or even seize an English port as a bargaining counter. The king's inability to decide what the fleet was for comes through very plainly—if such confusion may be said to be plain—in the letter in which his secretary informed the Duke of Medina Sidonia of the fleet's purpose in February 1587:

What the English have seized in Holland and Zealand, together with the fact that they infest the Indies and the seas, is such that it cannot all be recovered by defensive means alone, but rather obliges us to set fire to their own house, and that with so lively a flame as may force them to attend to it and withdraw from other theatres ... or at least, to take Ireland from them to serve as a prize to exchange for the places they hold in the Low Countries and so to quash the monster that devours the money and manhood of Spain, or as a means of undertaking the enterprise of England itself, if God does not open up some other way in the mean time.

Peace negotiations might have opened 'some other way'; but once the

Armada was taking shape and consuming money, the argument for using it became irresistible. As late as March 1588, when the momentum to launch the fleet was already irresistible, the Duke of Parma continued to urge the king to stay his hand. 'I should be failing in my duty,' he wrote,

if I did not inform Your Majesty that the general opinion is, that if the English proceed straightforwardly, as they profess to do, and their alarm at Your Majesty's armaments and great power compels them to incline to Your Majesty's interests, it would be better to conclude peace with them. By this means we should end the misery and calamity of these afflicted states [of the Low Countries], the Catholic religion would be established in them and your ancient dominion restored, besides which we should not jeopardise the Armada which Your Majesty has prepared, and we should escape the danger of some disaster.

This was good advice, as it turned out, but by then it was too late to heed it. Philip had decided that peace negotiations should be used only 'as a feint' unless and until the English conceded his essential demands of withdrawal from the Netherlands and the Indies and toleration for Catholics in England. Don Diego Pimentel, when he sailed with the Armada, was presumably typical of his comrades-in-arms in recognizing this. Asked by his Dutch captors 'if there might be made a good peace between England and Spain, he saith no, or very hardly: except that it were on such condition, that the king might so bridle the queen of England, that she should stir no more thereafter'.[4]

The Armada would thus certainly be used. The questions, however, of exactly where, and how, and for what purpose, remained open-ended. To understand the vagaries of Spanish strategy it is necessary to trace its slow evolution over a relatively long period of time, beginning when invasion plans directed against England were first mooted in earnest by the Marquess of Santa Cruz in August 1583. Flushed with victory in the Azores, the Marquess was in expansive mood:

Victories as thorough as those which God has deigned to confer on Your Majesty usually embolden princes to undertake new enterprises, and since Our Lord has made you so great a king it is right that you should enhance your latest triumph by ordering such preparations as are necessary to embark next year on the enterprise of England, for this will be a matter wholly of the Lord's service and good for the glory and utility of Your Majesty. Being so well armed and having such successful warriors, do not forego this chance.

The idea that the conquest of the Azores might portend the conquest of England was rather like the German hope in the last war that success in the Channel Islands or Crete might point towards the same end. The Marquess's other main argument was that the problem of the Low Countries could best be dealt with via England. At that stage, however, the commander on the spot in Flanders, the Duke of Parma, did not share the admiral's opinion. He felt strongly that the case was the other way round and that the distraction of a campaign elsewhere would put the Low Countries in jeopardy. English provocations, moreover, had not yet over-strained Philip's tolerance: in particular, there was no overt English expeditionary force in the Netherlands, so that Santa Cruz's arguments for tackling England first seemed weak. The king therefore gave the plan only a guarded welcome. 'These are matters', he told the admiral, 'of which nothing can be said for certain at the moment, since they depend on propitious timing and on chance occurrences which may create an opportunity in the future.' He ordered hard tack to be hoarded and galleons to be built, however. The project was shelved, but shelved within reach.[5]

By late 1585, changing circumstances had revived interest in Philip's entourage. In the first place, the nature of the entourage had itself changed. From October 1585, business was concentrated in the hands of a 'war cabinet' of five, later four, members, known as the *Junta de Noche* or Night Committee because it transacted business at the end of the day, referring decisions for the king to make before early mass the following morning. Its members were two courtiers of long administrative experience, Don Cristóbal de Moura and the Count of Chinchón, and two bureaucrats, Mateo Vázquez and Juan de Idiáquez, who were of modest origins and formed part of the 'aristocracy of service' created by the king's need of educated and energetic creatures to help him discharge the burden of paperwork. The Armada fell particularly within Idiáquez's sphere of responsibility, but all the important papers connected with it were seen by the king and Vázquez too. The *Junta de Noche* was an efficient machine for the conduct of business and genuinely increased the potential range of the monarchy's activities and the length of Philip's reach. In the winter of 1585, the king showed every sign of enhanced concern, even urgency, for the enterprise of England. He procured maps of the coasts, called for more espionage agents in England and Scotland, read chronicles of earlier invasions, and perused statistics of England's population and power. The importunities of English Catholics may have had some effect on Philip: at least their pleas that 'their co-religionists who remain in England should be delivered from the persecution they undergo' provided him with a useful pretext. But the exiles' chorus had been heard, without being heeded, many times before. If the king was now willing, as never before, to attend to their pleas, it was because the dictates of his own interests seemed to have changed. The decisive theatre, where events drove Philip to reappraise his attitude to England and to reactivate his invasion plans, was the Low Countries. In August 1585 Elizabeth took Philip's Dutch rebels under her protection in the Treaty of Nonsuch, pledging 5,000 foot and 1,000 horse to their war effort for as long as it lasted and appointing a representative to the rebels' supreme organ of state. The first contingent arrived within days, and in December the Earl of Leicester, the queen's acknowledged favourite and the most influential man in England, arrived to command them and to usurp rights of government, which, in Philip's estimation, belonged solely to the king of Spain. That Christmas, 'God Save the Queen!' was heard in Delft, Rotterdam, and Dort 'as if it had been in Cheapside'.[6]

Meanwhile, English intervention notwithstanding, the Duke of Parma's campaign against the Dutch had made remarkable progress. Parma had formerly made the mistake of telling Philip that an enterprise against England would have to await victories in the Netherlands and, in particular, the recovery of Antwerp. 'Now that—thanks be to God for it—that town is in my power,' Philip replied on 29 December 1585, 'it would suit me to know your suggestions for this business at once. For it is the only means to find a definitive remedy to put an end to the evils which [the English] prepare over there against the service of God and my own.' The king was no longer interested in Parma's views on the principle of an attempt against England: he was demanding practical plans to be drawn up for when the time was ready.

By the beginning of 1586, according to Santa Cruz, the time was favourable: the Turks preoccupied, France distracted, and Philip's power at its height. The king's calls for detailed proposals were answered all too volubly and soon four distinct plans were in circulation. The first, and most straightforward, was Santa Cruz's own. This called for a direct invasion of England from Spain in a massive, overwhelming force. The king found it 'very well set out, and it will be looked into for when the occasion arises'; yet he seems never seriously to have considered it. In the first place, it was far too cumbersome and costly, calling, for instance, by its author's computation, for more ships than were available in the entire Spanish monarchy. Secondly, it seemed wasteful to create an invasion force afresh in Spain, when a large concentration of seasoned veterans was already available, hard by England, in Flanders. The king had referred to Flanders as a possible base for the projected invasion in his earliest communication on the subject to Santa Cruz. In Madrid, Parma's plans were therefore eagerly awaited. When these arrived in March 1586, they were responsible for arousing wildly excessive expectations which blighted Spanish planning thereafter and may in the long run have been decisively

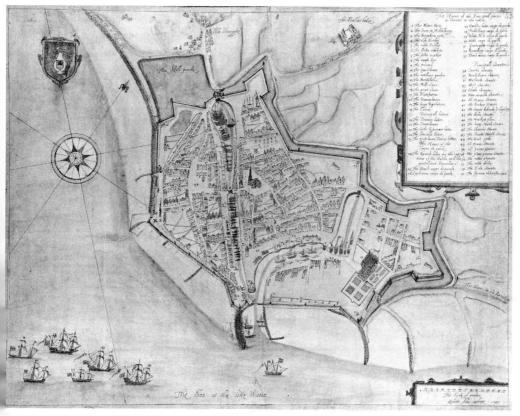

The Earl of Leicester's flotilla, straining eagerly in the breeze in the lower left-hand corner of the composition, arrives belatedly at Flushing in December 1585 to cement the Anglo-Dutch alliance and to exercise vigilance on the queen's behalf over the Dutch government. Leicester's stay did more to strain than strengthen the alliance. Notice the town's impressive fortifications and well-protected harbour. Philip II might have done well to heed his own early inclination and withhold the enterprise of England until Flushing was in Spanish hands providing a safe haven for the Armada's operations.

responsible for the errors on which the expedition foundered. The essential features of Parma's plan were these: (1) the invasion would be launched from the Netherlands; (2) the role of the Armada was to be diversionary; (3) Parma would cross in vessels of his own with a force of 30,000 foot and 500 light horse; (4) for this purpose he would build a fleet of transports with draft of two to three feet (against at least twenty required for a fighting ship) and concentrate them in the shallow-water harbour of Dunkirk; (5) the crossing could be effected in ten or twelve hours, or eight if the weather were exceptionally favourable; (6) provided the plan remained secure, secrecy would be ensured by the low profile of his flotilla, invisible until it neared England: in the absence of secrecy, the English fleet must be defeated or deflected. Parma's adviser, the engineer Piatti, later amplified this in Madrid: without cover of secrecy, the army would have to stay in Flanders; (7) he would disembark near the mouth of the Thames between Dover and Margate, or on the north bank if necessary, though here it was feared, as Piatti later explained, that the English cavalry might have an advantage; (8) resistance would be feeble: the people of those parts were unwarlike because they were rich and London was undefended. The only element of realism in this plan was represented by the stress laid on the need for secrecy—without which the supposed diversionary effect of the Armada, on which any chance of an unopposed crossing for Parma depended, would be out of the question. Yet even here fantasy obtruded: warlike preparations on the scale envisaged could hardly have escaped the notice of even so ramshackle a secret service as the English.[7]

Probably as a result of this plan, it seems to have been believed in Madrid, throughout the years in which the enterprise of England was planned and executed, that Parma would be able to cross the Channel in his fleet of barges, more or less independently of the Armada, if the sailing conditions were right and if the English navy was elsewhere. Piatti, indeed, assumed that the Armada would not arrive until 'a few days after' Parma's army disembarked. In fact, with Dutch marauders always on the coast, he would have been unlikely even to get out of harbour; and even if he had, the flower of Philip II's army would then have been wallowing, undefended and unescorted, on a notoriously treacherous sea. In any case, the vital pre-condition of secrecy could never be guaranteed. In no subsequent document is Parma anything like as confident: his sanguine expectations of good sailing conditions vanish; his need for direct help from the Armada becomes first explicit, then strident; his belief in the possibility of secrecy wilts. Why was he so reckless, or so deluded, in this first formulation of his plan? The plan has been attributed to the Catholic exile Hugh Owen, and may reflect the expatriates' typical world of wishful thinking. Or it has been suggested that Parma's real motives were concealed in the early part of the memorandum, which called for his forces to be brought up to strength and in Piatti's view that, for want of secrecy, the invasion force might have to stay in Flanders: by this interpretation Parma was exploiting the king's enthusiasm for the Armada project to extort reinforcements which could then be used in the Low Countries. It may be, too, that Parma's plans were like Pope Sixtus's ducats—promises he never expected to be called on to fulfil, because he assumed the Armada was a pipedream. To his own secretary, the historian Luis Cabrera de

Córdoba, he appeared even in the week before the Armada's arrival like a man who could not bring himself to believe that it was really on its way. Probably a large part of the explanation lies simply in Parma's need to put something on paper urgently, to satisfy the king. Philip's peremptory demand for plans had to be met and it was more important to offer something than to offer something sound.

Certainly, as soon as the plan was tendered, Parma's sense of urgency left him and his doubts began to show: on the very day it was dated he wrote to the king to say that the project needed further careful consideration and that he would send the engineer Piatti to discuss it in Madrid; this plea and promise were repeated two months later, with the justification that so great an affair demanded long reflection. In the interim, Parma confessed he had lost confidence: formerly, he said, he had considered the enterprise of England 'specially worthy of the greatness of his king' and had not regarded it as impossible. Now, however, a surprise attack could no longer be mounted; England would be prepared; the favourable time was past. 'If, despite all this,' he wrote, 'Your Majesty wishes to proceed, I hope he will do so with the prudence which characterises him.' In fact, Parma continued to correspond for the next fifteen months as if security had not been breached, but he never again assumed an encouraging tone.[8]

The remaining plans under consideration in Madrid called for the Armada to be directed to Scotland or Ireland. The 'Scottish' plan had its remote origins, like many of the hare-brained schemes that circulated at the Spanish court, in a plot by disaffected Catholics, who had appealed for Spanish help in 1582. It was easily dismissed by its opponents: the journey south from Scotland, they argued, was too long; the land had more heretics in it even than England; popular resistance would be sure to materialize; the country was poor in wheat and it would be impossible to forage for supplies; the army would be forced to march through a hostile environment of 'wilderness, marshland, mountains, woods and rivers'; Berwick was a formidable stronghold and the north of England was generally well defended; the English Catholics would not necessarily welcome the Scots, much less their doctrinally heterodox king; a plan that must work to the benefit of James VI could end by perpetuating heresy in England. Nevertheless, the version of the 'Scottish' plan which has survived among the papers of the Marquess of Cabra bequeathed two important legacies to Spanish strategical thinking on the Armada. It left the impression that some useful work might be accomplished in Scotland, if the Armada should be forced there by the weather. Moreover, it included a stipulation that became a vital part of all subsequent plans: despite the

The need for propaganda made Parma lavish with patronage of artists. Ott van Veen, the painter of this portrait, was a highly suitable choice for Parma's court, for he had been forced from Leyden to Antwerp by religious persecution at the age of sixteen. In 1585 he joined Parma's entourage, replacing Joos van Wingle who went into self-exile in Lutheran Frankfurt 'for the sake of his conscience'. Note the golden fleece on Parma's chain. The field breastplate he wears here should be compared with the parade armour on p. 100.

want of a deep-water harbour in Flanders, Parma's infantry could be embarked in Dunkirk or Nieuport 'when the main body of the fleet arrived from Spain'. The Scottish plan may have had some attraction in conjunction with Parma's plan as the source of the 'diversion' which would free the Channel for Parma's putative crossing—indeed, at one point, Piatti seems to have favoured it for that purpose; but it was hard to see how it could be guaranteed to distract the English fleet. It was more reasonable to suppose that Ireland, which was an English possession, would be defended at sea, and it may have been for that reason that the option of using the Armada against Ireland was more widely canvassed. Perhaps the most eloquent advocate of that plan was Sir William Stanley, dazzled by the hope of achieving in the Spanish service a conquest which had eluded the English army. He could, he boasted, conquer Ireland with 6,000 men. The seizure of Waterford or some other Irish port would provide Spain with the naval base she needed in northern waters, as well as posing a threat to England. It was at first in conjunction with Parma's plan that the Irish option seemed most likely to be adopted. In September 1586, however, Parma decided that Ireland was too far away to be the focus of an effective feint. Still, Ireland remained a complication that bedevilled Spanish strategic thinking: almost until the Armada sailed, it was assumed that Ireland would have some part in the Armada's operations, if only as a second front to be opened by the returning fleet after the invasion of England had been effected.[9]

The strategy adopted by the Armada followed none of the plans of the spring of 1586, but was an uneasy synthesis of all four. Its central feature—ironically, in common with the least influential, 'Scottish' plan—was for Parma's crossing to be synchronized with a covering operation by the Armada in the Channel. Even that principle seems to have been far from clear to some of those taking part; and it begged, in practice, a series of questions which were never answered, or else answered from different quarters in contradictory ways. The possibilities were limited, and the broad lines of strategy imposed, by circumstances. Two changes in particular, in the late summer of 1587, seem to have had a determining effect: first, the disappearance of any hope—for there had never really been any prospect—of an independent surprise attack by Parma; secondly, the new opportunity created at the beginning of August by Parma's capture of the port of Sluys—'the door', as royal instructions for the Marquess of Santa Cruz expressed it, 'that it would seem God has opened for us in Flanders, by our capture of Sluys, whence the passage to England is so short and secure'. Sluys was not the port that Spain ideally needed: there was no deep-water harbour there and the coasts of Holland and

Zealand, where the really useful ports were, remained firmly in enemy hands. Yet Sluys did make a material difference: Parma was now the incontestable master of the estuary of the Scheldt and could concentrate his invasion flotilla in safety and in strength.

In Madrid, the event seems to have been interpreted as favourable to a strategy that involved a 'junction' between Parma and the Armada; if so, the inference was a false one, for the Armada was too big, and its ships were too big, to anchor safely anywhere near Sluys. This fact, of course, was well known to Spanish decision-makers. Its significance, however, seems to have eluded them.

In any case, King Philip's advisers were being driven to the same conclusion by the forfeiture of the surprise element on which plans for an independent crossing by Parma depended. Spanish planning seems to have abandoned hope of surprise from early September, when the king wrote to Parma aborting the plan for an independent crossing on the grounds that it would leave Parma's force too exposed:

We have come to the conclusion that the plan of trying to avoid this difficulty by dividing our forces [*that is, presumably, using the Armada to draw off the English fleet*] and sending the Armada to attack some other point, might have an uncertain result. The enemy would understand the object of the manœuvre ... and they would concentrate their forces to oppose you instead of being diverted by the feint. I have become convinced, therefore, that the most advantageous way will be to join your forces there with ours at the same time; and when a junction is effected the affair will be simplified and the passage assured. The whole force can then be promptly applied to cutting off the root of the evil.

It was not, apparently, because of any breach of security that Spain had abandoned the strategy of surprise. On the contrary, Philip continued to suppose that his plans were a well-guarded secret. He assumed, in his letter to Parma, that at the time of the invasion the English ships would still be in port; he emphasized the need for speed—implicitly, to preserve secrecy. He stressed that dispatches should be secretly communicated and not even committed to writing. Parma's reply dissented from the plan, but accepted that security was still unbreached. In writing to Santa Cruz on 2 November, the king made the same assumption: 'It is of infinite importance that our main plan is kept secret ... and that care is taken that no destination should be credited other than Ireland—which still may perhaps in any case be the one we shall be obliged to adopt at the time.' At the end of the month, the king remained anxious to preserve security: 'you must not allow to leave that port any vessel that may take news or give information about you and the Armada to the enemy'. So secrecy, despite the high value Spaniards put on it, was not in itself decisive: plans for a surprise attack were dropped on practical grounds at a time when the Spaniards still believed in the effectiveness of their own security. The secret even had to be kept from two of the most senior naval officers—Miguel de Oquendo and Juan Martínez de Recalde—who had to be consulted about pilotage on the English coast 'but only revealing to them our intention to do something that will harm the English'.[10]

The effect of the loss of secrecy, when it came at last in November 1587, was not to change Spanish plans but to call them into question. The biggest gap in the leaky vessel of Spanish security was in Rome, where English fugitives would 'brag openly' and where 'Cardinals and great personages' paraded their knowledge of Spanish plans. English espionage was attuned to garnering such gossip—a bootless task for much of the time, which does not seem, for instance, to have yielded any reliable information about the Armada until the spring of 1588. But there were no resources for more sophisticated methods, much to the annoyance of commanders like Howard of Effingham, who complained that while Philip II knew exactly what was going on in England, Elizabeth should have spent 'but a thousand crowns to have had some intelligence'. There were good and bad spies on both sides: bad ones like Ascanio Cinnafarino, whose blunders and bad faith were such that Parma, embittered by experience of him, resolved to employ no more agents; good ones like Nicholas Oseley, who was rewarded in 1589 with the lease of a London parsonage 'in respect of his good service heretofore in Spain, in sending very good intelligence thence, and now since, in our late fight against the Spanish fleet' as 'one that hath so well deserved in adventuring his life in so many ways in her Majesty's service'. It sounds as if Oseley was a more active agent than most of those whose 'advices' clutter the archives; usually, they limited themselves to retailing rumour, which their masters, for want of anything better, heeded unduly to their frequent cost. The interrogation and torture of prisoners—harmless fishermen, for the most part, who knew less even than the spies—was another device much used to little profit. In the event, it seems not to have been through the usual channels that news of the Spanish plans reached London, though it was, indirectly, from Rome. Maurice of Nassau, entertaining his noble captive, Marzo Colonna, picked up all the conversation of the cardinals' tables and passed the relevant matter on to his English allies.[11]

With the secret out, Parma declared, success was impossible; the English and Dutch were forewarned and ready. This opinion, first voiced by Parma in late December 1587, as soon as he became aware of the breach of security, and frequently repeated, seems to have commanded little attention in Madrid. Parma professed himself 'bewildered' at his master's apparently abiding expectation that he could cross the Channel 'at once, if no enemy force impedes'. This sort of misunderstanding, this fatal inability to communicate clearly between Parma's camp and Madrid or the Escorial, never ceased, from now on, to dog the Spanish campaign. Parma repeated the message on 20 March in terms one might have thought unmistakable: 'if the enterprise were in the condition we had intended it to be, with respect to the vital point of secrecy, and so on, we might, with the help of God, look more confidently for a successful issue ... But things are not as we intended.' His pleas were ignored and, in a sense, he had only himself to blame, partly for having aroused false expectations with his initial contribution to the strategic planning process in the spring of 1586, partly for having introduced a new ambiguity, in letters of February 1588, by declaring himself powerless 'without the Armada to protect him'. This seems to have confirmed the impression in Madrid that the way to success lay in effecting a 'junction' between Parma and the Armada. There was a brief period—in or about January 1587— when the whole project looked as if it might be called off. In Florence, at least, a ripple of excitement was aroused by the hope that the Florentines' purloined galleon might be recovered from its Spanish sequestrators; and in England, rumours that the Spanish forces might be dissolved caused a certain amount of self-congratulation; 'it is the preparation that Her Majesty hath made', wrote Howard to Walsingham, 'that is the cause; for he cannot abide this heat that is provided for him.' In reality, however, the Armada was unstoppable, save by natural disaster or defeat. Spain had invested too much in the venture, not only in terms of material cost, which had to be justified by action, but also of aroused expectations and, above all, of the priceless 'prestige'—the royal *reputación*—which, rather than other more easily quantifiable forms of gain, was the object of Spanish policy. Parma's warnings were unheeded—the usual fate of a Cassandra's voice. Philip only urged the Armada on to greater speed.[12]

The deficiencies of Spanish planning have often been perceived as one of the principal reasons for the failure of the Armada. But English strategy was, for most of the time, in an almost equally helpless muddle. Ignorance of Spanish intentions was part of the reason for this and, though Howard was inclined to blame the English intelligence service, the fact that the Spaniards were uncertain of their own intentions may be held to exonerate the English of blame. Part of the trouble, however, lay in the fact that English decision-makers, like the Spanish, were divided between rival strategies, both on sea and on land, and havered and wavered between them.

On land the essential dilemma was whether the defensive forces should be spread around the coasts or concentrated at key points from which they could strike back against an invader. At one level, the choice rested on a straightforward military calculation: was it possible to prevent an enemy from landing? If not, the only remedy was to rely on effective counter-attack, which required the concentration of troops. In any case, of course, a more decisive result could be obtained by letting the enemy come ashore and then attacking him in strength with his back to the sea; but in the circumstances of the 1580s, safety was the English priority. Underlying the military controversy, however, was a conflict of rival social theories, already apparent in two works of 1579 which broached the issues that would be debated until, and only resolved in, the campaign against the Armada. Sir Thomas Digges's *Stratioticos* deplored the 'Home Guard' instinct that made every man rush to the shore to repel the invader in a disorderly rout, but believed that defensive forces had to be spread thinly about the coasts to sustain local morale. This argument was answered by Sir Thomas Wilford in *England's Defence*. Remorselessly pursuing the logic of Digges's own work, he argued that defence forces must be withdrawn to central rallying-points, from which they could concentrate rapidly against the invader and, in effect, pursue the advantage of a campaign on interior lines. Their difference of opinion was rooted in conflicting social prejudices. Both men favoured a 'citizen' army—the Renaissance ideal, advocated by Machiavelli—above a 'mercenary' force and subscribed to the common view, which the study of classical history was thought to justify, that the 'virtue' of the citizenry was enhanced by military service. Both agreed that the lowest orders of society failed to qualify. Digges, indeed, was particularly contemptuous: 'the fury of the country is great, every man (pro aris et focis) violently running down to the seaside ... A few orderly soldiers would close and pursue great numbers of these furious inflamed savage flock and herd.' He felt, none the less, that *pro aris et focis* was an effective slogan, capable of mobilizing men's resources of valour and energy in defence of their own homes, even at relatively low social levels. Though Wilford demurred, Digges represented the prevailing view. The bulk of the English land forces were organized at two levels: 'trained' bands, who corresponded roughly to the supposed antique ideal, drawn from the respectable classes and given periodic drill by professional soldiers, supplemented on paper, at least, by a large, heterogeneous, and ill-armed rabble of plebeian auxiliaries. The first category was capable, at the time of the Armada, of putting 26,000 men into the field—hardly enough to confront the sort of field force of over 20,000 seasoned veterans which the Armada might have landed;

and many, no doubt, justified Sir George Carey's jibe that they were 'trained rather in name than in deed'. The second category was computed at over 100,000 men more, but, even had they been levied—and they never were—it is hard to imagine them putting up a fight.[13]

The beacons which were intended to alert these forces represented all that was best and worst in the spirit of English defence. To the superficial glance of a Venetian visitor they seemed a thoroughly laudable institution:

Against all invasion they have this, to me, admirable arrangement. The whole countryside is diversified by charming hills, and from the summits of those which are nearer to the sea they sweep the whole horizon. On these summits are poles with braziers filled with inflammable material which is fired by the sentinel if armed ships of the enemy are sighted, and so in a moment the news spreads from hill to hill throughout the kingdom.

This was a succinct statement of the theory. The practice was more problematic. The beacons were maintained only for a limited season, locally determined but normally intended to run from spring to autumn. The cost and trouble were much resented: the extended season adopted in some southern counties caused 'murmuring' in Hampshire. In 1588 Lancashire maintained its watch only from 10 July until 30 September at a cost of 1s. 4d. per day. Throughout the country, the watchers were discharged on 6 October. The system was open to abuses alike of negligence and zeal. The justices of the peace were the appointed watchers of the watchers and did make periodic inspections of the beacons, distributing punishments for absenteeism, inattentiveness, drunkenness, and sleeping on duty; the watchers rotated, but a high incidence of neglect of duty was natural to the job. On the other hand, particularly at moments of international tension, the beacons intensified popular excitement, which in turn increased the hazards of a misfire by the nervous watchers. On the Isle of Wight the procedure was that if either of the two watches at the eastern and western extremities of the island saw between six and twenty ships, off the normal routes, heading for the coast, they would haul up a signal known as the 'ear'. Then one watcher would go to the nearest church and ring the alarm by clashing all the bells together: all the church bells of the island were silenced from 1583 onwards, except for single bells at Christmas, weddings, and funerals. Meanwhile, other watchers would warn the justices, or other persons appointed to verify the findings of the watch, who, if they confirmed the sighting of hostile sail, would in turn report to the Captain of the island. No beacon would yet be fired: that had to await the appearance of at least twenty sail, and the repeat of the entire procedure. Then one beacon would be lighted for twenty to thirty sail, two for thirty to fifty, and all three beacons for fifty or more. A minor beacon was presumably used for communication between the two watches. The effect of this system, when it worked, must have been to multiply false alarms and increase public fear, without providing the reliable early warning for which it had been designed. Its greatest usefulness was as a trigger of those popular sentiments of self-defence, *pro aris et focis*, admired by Digges, and as a means of making the common people identify with the system.[14]

The basic unit of organization was the county: this meant that without a decisive policy initiative at the centre the strategy of concentrated defence would be impossible, for the forces were necessarily spread throughout the variegated historic communities that supplied them. Only the 'Queen's Guard'—a 'professional' force nominally of 10,000 men—would be available as a central reserve at Westminster. By a miscalculation of Parma's intentions, the biggest concentration was to be in Essex, where it would be unable to meet a Spanish landing. In April 1588, however, English policy was reversed, or at least redefined, somewhat more in keeping with the doctrines of Sir Thomas Wilford. A commission of three seasoned soldiers—Sir John Norris, Sir Thomas Leighton, and Sir Thomas Morgan—was appointed to overall direction of the trained bands. The

The construction of the beacons is apparent in this plan of the beacon network around Portland. Note the merchant galley lying peacefully off the coast in the top right-hand corner.

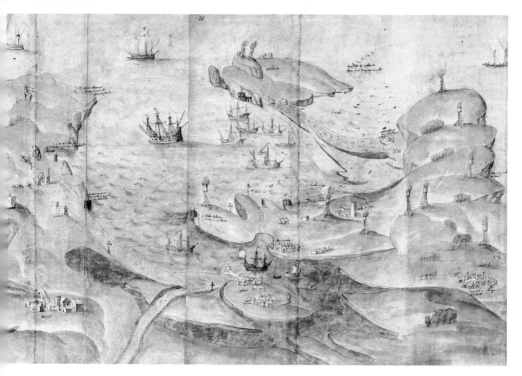

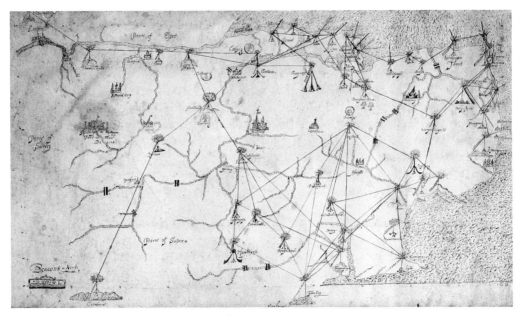

Map of the beacons of Kent made by W. Lambarde in 1585, showing their relative heights and the sight-lines linking them.

language of Leighton's instructions to the levies of Norfolk reveals a preparation thoroughly grounded in the doctrines of Wilford and Digges: 'It is to be doubted [i.e. *expected*] that upon any alarm or firing of beacon, the multitude will assemble themselves, in some disordered sorte . . . upon the rudeness of the people.' He insisted on the deployment of only selected and effective soldiery. He and his colleagues reallocated the various county levies into groups and ordered them to concentrate at points of maximum sensitivity; there was still no attempt to exploit a strategy of interior lines. Perhaps for Digges's reasons of morale, it was still thought advisable to be seen to be concentrating on defence of the coasts, and the bands were ordered to places thought particularly vulnerable to a Spanish landing, such as Falmouth, Plymouth, Poole, and Portsmouth. This strategy was, of course, as mistaken as that which it replaced; in some ways, it was worse, because it deflected attention towards the westward ports, where a direct attack from Spain might have been expected; but Spanish policy had implicitly ruled out such an attack at least two years before, and even English intelligence had known this for some months. Nevertheless, the strategy of massing relatively large numbers of men can at least be said to have had some purposefulness and professionalism about it.[15]

At sea, too, English strategy was almost crippled by controversy. Although Philip II seems to have believed that the English might choose to let the Armada disembark its invasion force and then attack a fleet depleted of soldiers, there was certainly never any such suggestion made in England. Everyone agreed that the Spanish should be 'impeached' at sea. The question was, how far out? Drake was the consistent advocate of an aggressive response. He had carried out a spectacularly successful raid against Cadiz and Sagres in April 1587: until he succumbed, as usual, to the temptation to go running after prizes, he managed to damage, though not to wreck, the Spanish war effort. He went on longing to repeat the trick and, as he hoped, wreak irremediable harm on the Armada. On 9 April 1588—a couple of days short of the anniversary of his Cadiz raid—he addressed a forceful plea to the Council, urging that 'with fifty sail of shipping we shall do more good upon their own coast, than a great many more will do here at home'. His principle of action was 'to seek God's enemies and Her Majesty's where they may be found, for the Lord is on our side, whereby we may assure ourselves our numbers are greater than theirs'. This proposed strategy was endorsed in a surprising quarter: Philip II had been expecting Drake to bring a third of England's fleet to Spanish shores since February; and when the Armada was under sail, its commander was ordered not to let a descent by Drake in the Spanish rear upon an undefended Spain allow him to hesitate in his main task. Spanish awareness of the possibility perhaps suggests that Drake's enthusiasm for a pre-emptive strike was misguided. A stratagem that had worked a year before would now encounter a well-prepared enemy.[16]

Of great concern to Elizabeth and her advisers was the fear that the fleets would pass at sea, leaving England denuded of defence. Even some of the professional naval commanders were reluctant to believe that the Armada could be forestalled: Hawkins was initially doubtful of Drake's views. Six ships, he thought, should permanently patrol the Spanish coast for intelligence gathering; for the rest, however, the queen's navy should have 'as little to do in foreign countries as may be but of mere necessity, for that breedeth a great charge and no good at all'. Drake and his kinsman Sir Thomas Fenner, however, continued to lobby for a policy of attack, especially when the June storm dispersed the Armada to various Spanish ports and seemed to make it more vulnerable. 'To seek the place of their dispersed companies and to seek by all possible means their waste' was Drake's purpose, and the storm seems to have convinced the other naval commanders that he was right. On 24 June, for instance, Howard wrote to Walsingham:

The opinion of Sir Francis Drake, Mr Hawkins, Mr Frobisher, and others that be men of greatest judgement [and] experience, as also my own concurring with

As in the well-known version by Marcus Gheeraerts the younger from Buckland Abbey, Drake's hand rests on a globe in allusion to his circumnavigation. The expert face-painting contrasts with the awkward figure-painting by a cruder hand. The globe is accurately drawn, in contrast to the shadowy world which rests under Elizabeth's hand on p. 240. Drake's palm is over the Indian Ocean, which, more than any other sea, was an almost unchallenged Luso-Castilian lake. Drake seems, therefore, to be making at once a gesture of pride in his own achievement in penetrating those waters and a claim to the freedom of the seas.

them in the same, is that [the] surest way to meet with the Spanish fleet is upon their own [coast], or in any harbour of their own, and there to defeat them.

Even in the face of this uncommon professional unanimity, the decision-makers in London wavered. Howard's proposals were imperfectly understood, as one of the official replies demonstrates: Her Majesty, wrote Walsingham,

> perceiving ... that you were minded to repair to the Isle of Bayona [off Cape Finisterre], if the wind serve, there to abide the Spanish fleet or to discover what course they meant to take, doubting that, in case your Lordship should put over too far, the said fleet may take some other way, whereby they may escape your Lordship, as by bending their course westward to the altitude of fifty degrees, and then to shoot over to this realm, hath therefore willed me to let your Lordship understand that she thinketh it not convenient that your Lordship should go so far to the south as the said Isles of Bayona, but should ply up and down in some indifferent place between the coast of Spain and this realm, so as you may be able to answer any attempt that the said fleet shall make either against this realm, Ireland or Scotland.[17]

This reply shows a number of disturbing things about English policy: that it was still not abreast of its own intelligence; that it was made by landlubbers who thought fleets could manœuvre like armies, without respect to wind and weather; and that communications between the capital and the front were as bad—as conducive, that is, to mutual misunderstanding—as those of Spain. Howard's rejoinder was withering: he explained that the queen's wish was 'debated' by Drake, Hawkins, Frobisher, Fenner, and himself. Affecting proper humility in rebuking his queen, Howard listed nine reasons for disregarding her command. The Spaniards were 'long victualled' and able to wait, if it suited them, for the English to consume their own rations; it would be unwise to leave the initiative in Spanish hands as 'the king of Spain will have all things perfect, [as] his plot is laid, before he will proceed to execute'; delay would give time for Spanish diplomacy and intrigue to bring France into the war against England; to wait 'on and off betwixt England and Spain' would mean putting the English fleet to leeward of the Armada, and it was an invariable axiom of warfare in the age of sail that the advantage in battle lay to windward; an attack would put the English 'on the Spaniards' backs'; it would be impossible to get water in mid-ocean; experience showed that the fleet would not be able to 'ply up and down' for long; the queen's strategy would expose the fleet to storms; the queen herself, Howard finally recalled, had recently favoured an attack on Lisbon. Howard had formerly counted himself among the doubters, but had been

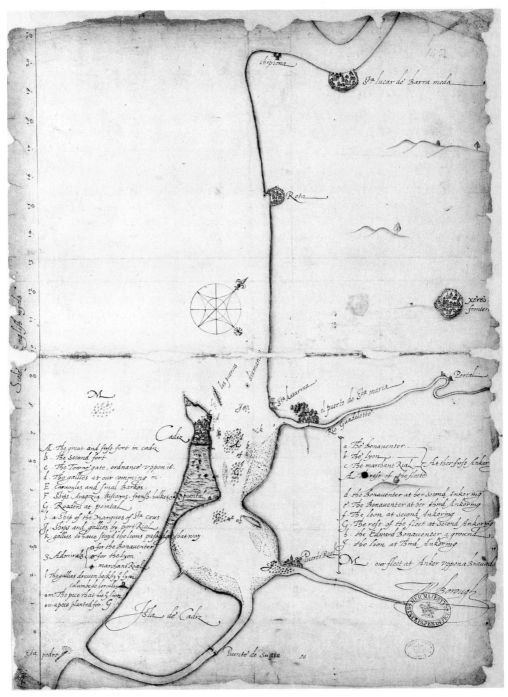

A sketch of the Bay of Cadiz and its environs made by William Borough at the time of Drake's raid in 1587. The positions of leading ships at various times, as well as of the prizes taken, the Spanish defending ships, and the shore guns are all marked with a key. Item *h* in the inner harbour is the galleon belonging to the Marquess of Santa Cruz, which was Drake's main objective and for which he risked his fleet before the Spaniards fired her. Borough was arrested for insubordination, having first tried to call Drake to account to his council of war and then retiring from the action on the second day's fighting.

well persuaded to change his mind. 'I did and will ever yield unto them of greater experience.' It was, in short, both impracticable and imprudent to dawdle at sea in the way the queen suggested. 'It is a hard matter and a thing impossible for us', wrote Howard in a further dispatch, 'to lie in any place or to be anywhere to guard England, Ireland and Scotland.' As usual, Howard blamed defective intelligence: 'I would to God Her Majesty had [thought] well for it that she had understood their plot, which would have been done easily for money.' Yet the fault lay less in English espionage—weak though that was—than in English inability to interpret their intelligence with confidence and take firm decisions accordingly.[18]

Howard won his struggle for aggressive orders. Ironically, however, no sooner had he obtained permission to attack the Armada in its home waters, than the new strategy was reversed again, this time by the weather. The wind that brought the Armada to England in late July penned the English in port. 'Sir,' wrote Howard to Walsingham,

I protest before God I wish I had not a foot of land in England, that the wind would serve us to be abroad ... and that it would have been easy matter to have made him not able to have troubled [Her] Majesty again in one seven years. And if the wind had been favourable unto us, we had been [so] long since before their doors [that] they should have not stirred but we would have been upon their jacks.[19]

Thus, had English hesitation been a little less protracted, or the wind a little less capricious, the Armada campaign might have seen the start of what has been the dominant tradition in English naval strategy almost ever since: the device of destroying or blockading an enemy in order to deny him any opportunity to attempt an invasion. As it was, the English allowed the enemy the initiative and awaited him on their own shores. At least this was better than cruising far off for prizes, which the English captains would have done had they followed their usual practice and, perhaps, their usual inclinations. In May 1587, Drake broke off his raid and patrol of the Iberian coast to recoup the cost by prize-hunting. Twice during the battle against the Armada the effectiveness of the English pursuit of the enemy was threatened by prize-seeking. At the time of the Armada there were strategists in England who would have elevated prize-seeking on the high seas into a general plan of war against Spain. The motive for it may have been piratical cupidity; the reason alleged, however, was that it would cut the veins that bore Spain's life-blood, the Indies trade-routes that carried bullion to Spain. Only two days before the Armada finally set sail for England, Thomas Cely was advocating a 'secret' strategy that seems to have involved removing England's naval defences to the Spanish sealanes. 'Do not think', he wrote, 'to have any quietness with the king of Spain as long as his moneys comes out of the Indies ... Within one year after it is done, it will bring Her Majesty to more quietness, and her countries, than all her Council in seven years.' Such an attempt would have been disastrous for England. Not even an attack on the Spanish Main would have deflected the Armada: that much, at least, was clear in Medina Sidonia's instructions. The bullion, it was true, might never be so vulnerable again; at Medina Sidonia's own suggestion, it was divided among 'swift, small ships' for the 1588 sailing, so that the Indies galleons could sail with the Armada; but it was hardly worth stripping England's home defences for.[20]

Five

Strategy: Stagnation and Extemporization

IN the winter of 1587, Philip II was growing impatient. It seems unlikely that the Spaniards can genuinely have contemplated a winter campaign: Parma's putative crossing in flat barges presupposed a tranquil sea; and not even the Armada, owing to its large number of Mediterranean ships, could face the northern waves without misgivings. As the king admitted, 'Great risks are involved in the moving of a mighty Armada in winter, particularly in that Channel with no port secured.' Yet the king hardly ceased to importune the Marquess of Santa Cruz to hurry, even at the cost of braving winter storms: 'the other reasons which have induced His Majesty to take this resolution are even weightier; and it is to be hoped that God, Whose cause this is, will give of His bounty suitable weather'. The king justified this shaky reasoning on the grounds that the enemy was unprepared. Yet he must have been aware that a winter foray was neither possible nor desirable. His promptings were directed rather at the lassitude of the Marquess of Santa Cruz than at any real prospect of an early campaign.[1]

For Santa Cruz no longer evinced any of the brio and verve with which he had first advocated the enterprise of England five years before. He had lost his enthusiasm for it and probably his confidence in it. He was ageing, weary, and sick. On 12 December 1587 he promised he would be ready within a month. Yet a month later, the king's vexation showed in a polite rebuke, elicited by Santa Cruz's report on the state of the fleet, for which he received thanks 'although we cannot see therein what we wanted to know, which is the exact day on which the Armada will be able to sail for certain, as we have asked you many times without being able to get any reply from you'. Towards the end of January, the king decided to send an inspector to oversee Santa Cruz's arrangements, 'although I am certain of your care and zeal which you show in my service and that you perform and accomplish it with all your customary attentiveness'. When the Count of Fuentes arrived in Lisbon to carry out his inspection, it was apparent from the first that he was dealing with a broken man: 'Three times he

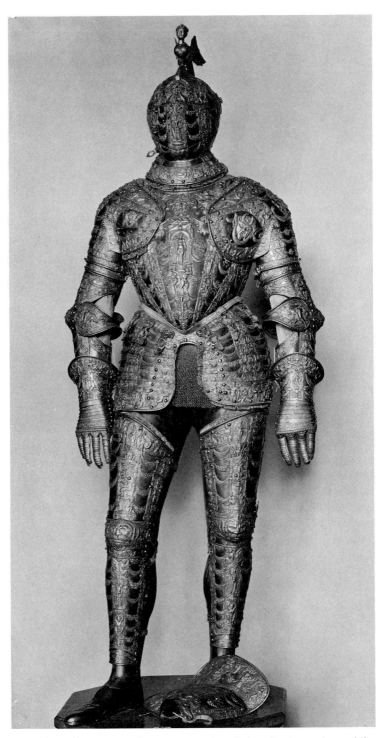

Parma's exuberant parade armour evokes the opulent grandeur of the monarchy the English were confronting. Alexander is depicted in classical armour on the helm and greaves and naked, leaning on a scimitar, on the cuirass. Victory forms the crest and appears on the spandrels. Parma, whose name was Alessandro Farnese, was hailed by contemporaries as the new Alexander, and as the new Hercules (cf. p. 131).

cast his eyes over the final lines of my warrant—I know not to what end. I found him in his bed and they tell me it is no wonder, considering how he toiled these last seven or eight days.'

Fuentes, however, seems to have been· overawed by his task. He was evidently not much of an observer and relied on what he was told. He was inclined to defer to the gallant old admiral and accept his own valuation of the state of readiness of the fleet. He did notice that the ships were some six or seven hundred mariners short and that the Squadron of the Levant—composed of the Mediterranean great ships—was badly in need of bronze ordnance. In general, however, he was unwilling to blame Santa Cruz for lack of competence. The delay was caused by the admiral's sickness, the want of a fair wind, and the need to embark more money for future pay and disbursements *en route* before the fleet should depart. Once these deficiencies were remedied, all would be well.[2]

The first of them, however, was irremediable. On 9 February, Fuentes reported that the marquess was dying. Later that day he recorded his death 'with which all things here are left in grave need of a master who understands them and knows how to manage them'. By the time it had been decided in Madrid that the Duke of Medina Sidonia should be appointed to command the Armada 'in view of the marquess's sickness', Santa Cruz had been dead for two days.

Well known though it is, Medina Sidonia's response to his preferment will bear any amount of requotation:

I first humbly thank His Majesty for having thought of me for so great a task, and I wish I possessed the talents and strength necessary for it. But, Sir, I have not health for the sea, for I know by the small experience that I have had afloat that I will soon become seasick and have many humours. Besides this, your Worship knows, as I have often told you verbally and in writing, that I am in great need, so much so that when I have had to go to Madrid I have been obliged to borrow money for the journey. My house owes 900,000 ducats, and I am therefore quite unable to accept the command. I have not a single *real* I can spend on the expedition. Apart from this, neither my conscience nor my duty will allow me to take this service upon me. The force is so great, and the undertaking so important, that it would not be right for a person like myself, possessing no experience of seafaring or of war, to take charge of it. So, Sir, in the interest of His Majesty's service, and for the love I bear him, I submit to you, for communication to him, that I possess neither aptitude, ability, health nor fortune for the expedition. The lack of any one of these qualities would be sufficient to excuse me, and much more the lack of them all, as is the case with me at present. But besides all this, for me to take charge of the Armada afresh, without the slightest knowledge of it, of the persons who are taking part in it, of the objects in view, of the intelligence from England, without any acquaintance with the ports there, or of the arrangements which the Marquess has been making for years past, would be simply groping in the dark, even if I had experience, seeing that I should suddenly, and without preparation, enter a new career. So, Sir, you see that my reasons for declining are so strong and convincing in His Majesty's own interests, that I cannot attempt a task of which I have no doubt I should give a bad account. I should be travelling in the dark and should have to be guided by the opinions of others, of whose good or bad qualities I know nothing, and which of them might seek to deceive and ruin me. His Majesty has other subjects who can serve him in this matter, with the necessary experience; and if it depended upon me I should confer the command on the Adelantado of Castile, with the assistance of the same counsellors as are attached to the Marquess. He would be able to take the fleet from here [*that is, Cadiz, where Medina Sidonia was in charge of assembling a squadron for the Armada*] and join that at Lisbon; and I am certain that the Adelantado would have the help of God for he is a very good Christian, and a just man, besides which he has great knowledge of the sea and has seen naval warfare, in addition to his great experience on land. This is all I can reply to your first letter. I do so with all frankness and truth, as befits me; and I have no doubt that His Majesty, in his magnanimity, will do me the favour which I humbly beg, and will not entrust to me a task of which, certainly, I should not give a good account; for I do not understand it, know nothing about it, have no health for the sea, and no money to spend upon it.[3]

This was no conventional *nolo episcopari,* though the implication that the writer's moral character would jeopardize the expedition's chances of securing divine goodwill owes something to a long tradition of modest disclaimers. What is remarkable about the letter is the strength of Medina Sidonia's resistance. His unwillingness rings with truth. Yet much of what he said was obviously disingenuous. It was, for example, precisely for his wealth, among other things, that he was chosen for his role; the debts he mentions represented a small charge on the most extensive aristocratic patrimony in Spain; and, in the event, he did spend heavily out of his own pocket in the course of his command. Though it may be true that he was prone to seasickness, his general health was robust and, without sparing himself toil or danger, he endured the Armada voyage better than many more seasoned subordinates. The king acknowledged the duke's habitual indifference to his own physical well-being, urging him, for instance, on 30 May 1588, 'to look to your own health, of which I am told you take little care'. This plea was to be repeated in almost the same words at the conclusion of the campaign. Moreover, Medina Sidonia's claim to know 'nothing' of the Armada was obviously false. Indeed, his letter of rejection of the command goes on to make arrangements for the dispatch of two squadrons to join the fleet and to offer recommendations about the use of galleys. Second only to Santa Cruz himself, Medina

Sidonia had been the most important figure in the making of the Armada. He had been privy to the king's plans at a time when the main subordinate commanders, who were actually to sail with the expedition, were not. Under the duke's aegis, in the harbour of Cadiz, supplies and ships were marshalled on their way to Lisbon. In and across Andalusia, under his control, much of the provender and most of the troops were mustered or marched.[4]

The excessive protestations of the duke's letter point to two of the king's main reasons for appointing him to the command: his wealth, which the king hoped to exploit in royal service; and his intimate knowledge of and involvement in the Armada, which he had discharged with notable efficiency. Other reasons of no less moment seemed to urge the same course. The duke was the model of a form of aristocratic propriety that was entirely Spanish—a brave bullfighter with perfect manners and turned-out toes, as reported by Juan de Vitoria, 'prudent, brave and of extreme generosity' as recorded by Lippomani. Even by his enemies these virtues were acknowledged. Although, judged by most standards, he was the greatest magnate in Spain, there was no whisper of suspicion of his loyalty to the crown: indeed, it is much to the credit of Philip II's government that in the enterprise of the Armada he was served, reluctantly and against their better judgement, yet selflessly and with unstinting obedience, by two great territorial princes—Parma and Medina Sidonia— who, in almost any other European monarchy at the time, would have seemed best qualified to defy rather than to serve the monarchy. By virtue of the greatness of his lineage, and the propriety of his personal qualities, the duke commanded, as the king's secretary observed, the very 'prestige and good opinion of the world' that the commander of a great fleet, from an emulous and heterogeneous monarchy, needed in abundance. In particular, the people of Andalusia, the region on which the Armada critically depended for supplies and men, owed the duke affection and allegiance. The Venetian ambassador predicted that he would be followed to England by 'many nobles and all Andalusia'; the king believed that men would be more disposed to serve in the Armada, and in greater numbers, if the duke took up the command.[5]

In claiming to have no martial expertise, Medina Sidonia was evidently strictly truthful. But that was not what the king wanted, or thought he wanted, at the time. Reading between the lines of the Count of Fuentes's reports, the members of the *Junta de Noche* must have realized that Santa Cruz had left the fleet in a shambles; what was required was good management, not expansive panache, a trusted administrator rather than a proven field commander, a man who would sacrifice initiative to obedience and *élan* to economy. All these requirements seemed, directly and rightly, to indicate the Duke of Medina Sidonia. His 'good management of my property', his 'excellent management and vigilance' were the qualities praised by the king; to Oquendo he displayed 'a good grasp of everything and very good foresight'; a passage of the king's instructions to the new commander reveals exactly why Medina Sidonia enjoyed royal trust:

The experience I have had of your constant efforts towards the economy of my treasury gives me great hope that in all matters of expenditure connected with the Armada, you will spare as much as possible the money you are carrying with you in the fleet. You know how much trouble it has cost to collect it, and the necessity from which we are suffering, and you will take to heart the care of seeing that the musters are made with great precision, and that no trick is played upon you with regard to the number of men. This is not only a question of expenditure but very often of success or failure. You will not forget to take particular account of the quality of the victuals, and their good preservation and distribution, so that they may not be exhausted or run short before the time, as the health and maintenance of the men depend so much upon this. You will keep your eyes constantly on the officers of all branches of the service, so that your vigilance may stimulate theirs and thus that every man on the fleet may be kept on the alert to do his duty.

These were precisely the respects in which Santa Cruz, in his final illness, could be seen to have failed.[6]

While Medina Sidonia's appointment was 'extremely well praised' at the time—and there is no reason to dismiss this as royal flattery of the new commander, for the Venetian ambassador confirmed it—some contemporary misgivings can be detected. The French ambassador thought it showed the dearth of suitable candidates from whom the king had to choose, the 'lack of ministers for the conduct of affairs'. Juan Martínez de Recalde, who recommended himself for the job, thought the commander of the Armada should be a veteran admiral; and Medina Sidonia himself evidently concurred. On the whole, however, the chorus of praise was remarkably harmonious. Some of the stories told against the duke were manifestly thought up with hindsight once the failure of the Armada had served to make men wise after the event. The most outrageous of these, told by Juan de Vitoria, depicts the distress of the duchess on hearing of her husband's new command. 'I know the duke is good for inside his own home,' she is supposed to have told her ladies between her tears, 'and for where men do not know him.' This sort of evidence is not to be trusted, and it seems safe on the whole to conclude that Medina Sidonia's appointment was a wise one, well thought of and justified by the positive qualities the new commander brought to his task.

After the Armada Medina Sidonia lived reclusively, and portraits of him are rare. The austerity of this portrait exceeds that of others, even from Philip II's puritanical court. Apart from the collar of the Golden Fleece, the duke wears no ornament. His simple neckband contrasts with the ruffs worn by the subjects of other portraits (see pp. 84, 94).

On the other hand, there were three respects in which the king's choice contributed to the stagnation of Spanish strategy, which was to characterize most of the campaign, and therefore, perhaps, to contribute to the unsuccessful outcome. In the first place, the Armada was now in the hands of two men—Parma and Medina Sidonia—neither of whom had any confidence in its mission. This was the real reason for Medina Sidonia's desire to forego the command; the mixture of sincerity and mendacity which ran through his letter of regret arose from the fact that he could not be explicit about his conviction that the enterprise was hopeless. He concealed his disbelief in the expedition under protestations of disbelief in himself. Other sources leave no scope to doubt the real ground of his diffidence: he never let slip any opportunity to advise the king to abandon the enterprise; dissembling, he repeatedly implied that the task was humanly impossible by representing the problems as a trial of faith; and, while doing his utmost to prepare the expedition for success, he did his best to prepare the king for failure. From the time of Medina Sidonia's appointment, the Armada was led in a fatalistic spirit, foredoomed by self-fulfilling prophecies of disaster.

Moreover, Medina Sidonia's lack of campaign experience, especially at sea, made the Armada a hostage to two crippling influences: rule by committee, and rigid adherence to an impossible plan. Philip demanded uncompromising obedience to orders, and the duke's reliability in this respect had been one of his greatest recommendations. When, for instance, the king heard that the Armada's pilots favoured an anchorage on the southern coast of England—which, as events turned out, might have offered the best hope of success—he informed Medina Sidonia that 'I am not anxious about this, as they are unaware of the order you have to proceed directly to the Cape of Margate'. This was a typical experience; initiative was discouraged, and Medina Sidonia's personality and brief helped keep it tightly reined, to the Armada's cost. Decisions which orders from Madrid could not anticipate, including the tactical direction of battles, had to be left to councils of war and to the professional advice with which, by royal command, the duke was provided. His naval adviser, Diego Flores de Valdés, was in the unenviable position of bearing responsibility without enjoying power, pre-cast in the role of scapegoat. More even than Medina Sidonia he became the victim of hindsight: the only Armada commander to be permanently and officially disgraced. It was the 'common opinion', according to a report commissioned by the king, 'that Diego Flores has performed his office badly'. He was accused of treachery by his cousin, Don Pedro, and of cowardice by the unquestionably brave Don Francisco de Bobadilla. When Juan Martínez de Recalde returned from the campaign, near to death with fatigue, he was 'inconsolable to see how so glorious a victory has slipped from our grasp, and all because Oquendo was not appointed to advise the Duke, as I recommended here and on the voyage, for I was not satisfied with the person the duke had chosen'.[7]

In fact, Diego Flores was as much the king's choice as the duke's and had been selected for the same reason as Medina Sidonia himself. Philip wanted no displays of initiative but solid adherence to the plan dictated in Madrid; this was not only because of the Spanish policy-makers' inflexibility of taste, but for the good reason that the plan depended on cooperation between the widely separated forces of the Armada and the Army of Flanders, neither of whom could be relied upon to cope with unexpected initiatives by the other. Diego Flores behaved throughout like a royal watchdog: all his advice was directed towards the ends of the plan: avoidance of risk, steady progress up the Channel, unflinching pursuit of the illusory 'junction' the Armada was supposed to effect with Parma.

Thus, against the consensus in the fleet, he advised abandonment of his cousin's ship when it straggled; he wanted to return to the Channel when the wind and sailing conditions had forced the fleet into the North Sea; and advocated a return to the Channel rather than a retreat to Spain, after the fighting, presumably in the hope of retrieving the lost possibility of co-operation with the Army of Flanders. If Oquendo had been selected for the role, his advice to the duke would always have been opportunistic and impetuous, far from what the king required. Oquendo, for example, was fretting to leave Lisbon in early May and wanted to be allowed to lead an armed reconnaissance to England ahead of the fleet; when the Armada was penned up by the weather in Corunna in July, Oquendo was the first to want to be off, moving his squadron out into the roads and urging his chief to do the same. And at the first sight of the English fleet, Oquendo, with most of the other veterans and, as things turned out, probably rightly, called for an immediate attack. By the combination of Medina Sidonia and Diego Flores in the position of supreme responsibility, such impetuousness was bound always to be restrained in 'prudence'.

It was never intended that Diego Flores should monopolize access to the duke. Francisco de Bobadilla was at the commander's side to give advice on matters of soldiering, though, in the event, it was always nautical problems that were dominant. All strategic and major tactical decisions had to be referred to a council of war, which brought together the squadron commanders and some others; but it is in the nature of councils of war to be divided, and with a commander like Medina Sidonia, the effect of the system was to contribute further to the stagnation of strategy. Philip II was aware of the mutual jealousies which could bedevil such councils; and Juan Martínez de Recalde accurately identified its effects when he pleaded, during the wait in Corunna, that 'for this service to turn out well, it is essential that bad blood should cease among those who hold commands, or at least that ill performed duty should cease, which does nothing save to cast the blame elsewhere.'

When the campaign was over, Andrés de Alba, who was in charge of co-ordinating relief for the stricken fleet, took the view that the 'discord within the Armada caused by the commanders themselves was greater than that inflicted by the enemy'. Don Pedro de Valdés claimed that it was his independent-mindedness in council that turned Medina Sidonia against him; after he advocated an immediate departure from Corunna on 27 June 1588, regardless of the storm which had scattered the Armada, the duke 'looked on him with an unfriendly eye and had used expressions towards him which grieved him'. This impression may have been the result rather of Don Pedro's imagination or excessive sensitivity than of

Medina Sidonia's intention: unfailing courtesy towards subordinates was the duke's usual rule, typical of his spirit of *noblesse oblige*. Certainly, however, Spanish councils of war worked poorly, neither contributing effectively to decision-making nor enhancing morale. The first case of the duke's accepting the advice of a council along lines not laid down in his instructions occurred after the battle was over, when the decision adopted—to return to Spain—was made inevitable by circumstances, and was clearly, if resignedly, espoused and advocated by the duke himself.[8]

The English case, on the other hand, demonstrates that a decision-making system based on the council of war could work well. This was not because the English command was less riven by jealousy and animosity than the Spanish. England had 'men at variance' enough. Frobisher's hatred of Drake and Drake's rivalry with Howard were potentially

It did not take long for Medina Sidonia's orders of 28 May 1588 to reach English hands. They had been translated and printed in this edition, enhanced by a rare woodcut of a galleon in full sail, by August. William Byrd had a copy among his many books of Spanish interest or provenance. Though the translator is given on the title-page as an unidentifiable 'T.P.', the initials 'Th. N.' appear at the end of the text— possibly a reference to the Thomas Nash who wrote *Piers Penniless* (see p. 238).

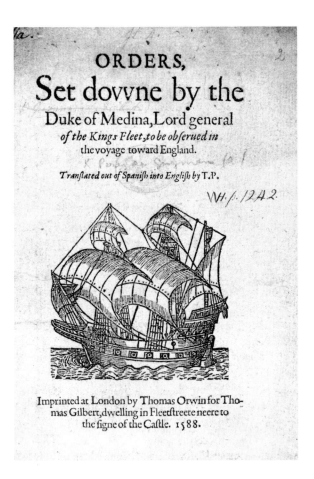

ORDERS,

Set dovvne by the

Duke of Medina, Lord general

of the Kings Fleet, to be obſerued in
the voyage toward England.

Tranſlated out of Spaniſh into Engliſh by T.P.

WH./. 1242.

Imprinted at London by Thomas Orwin for Thomas Gilbert, dwelling in Fleetſtreete neere to the ſigne of the Caſtle. 1588.

highly disruptive. Drake felt qualified to be commander-in-chief: the narrative of the Armada written under Drake's patronage by the Italian Protestant propagandist Ubaldini claims that it was only because it was necessary to match Medina Sidonia, 'a prince of consequence', that he was passed over in favour of the aristocratic Howard. Yet despite the loss of patronage Drake suffered in consequence, despite the personal expense to which he was put by the late arrival of supplies, and despite his pride in a reputation as 'fatal to the Spaniards', Drake seems to have managed to limit his display of his resentment to complaints 'leaked' through Ubaldini. The Italian alleged it; Howard himself confirmed it. According to Ubaldini, though 'a man born and brought up among freebooters would have found it irksome to practise the self-restraint admired by the ancient Romans', Drake showed Howard every form of deference and made himself 'of one mind and thought with the Admiral'. Howard was so gratified by Drake's dealing with him that he told Walsingham, 'Sir, I must not omit to let you know how lovingly and kindly Sir Francis Drake beareth himself; and also how dutifully to Her Majesty's service and unto me, being in the place I am in; which I pray you he may receive thanks for, by some private letter from you.' This amity did not survive the campaign unscathed; Ubaldini produced a second version of his narrative, much less favourable to Drake, probably with encouragement from Howard and 'other honoured men who served in the fleet.' Frobisher and Drake came almost to blows, with Frobisher threatening to 'spend the best blood in my belly' in furtherance of his complaints. Yet, while the Spanish Council was never able to show any flexibility or initiative, the English was capable of constructive decision-making. The outstanding example of this was the decision of 3 August 1588 to reorganize the English fleet into squadrons, apparently in imitation of the Spanish system and in acknowledgement of the impressive order in which the Spaniards kept rank and manœuvred concertedly.[9]

Not that English strategy was immune from stagnation. By the time the Armada arrived, the English sea commanders had extemporized a plan of sorts: 'so to course the enemy', as Howard put it, 'as that they shall have no leisure to land' and 'to follow and pursue the Spanish fleet until we have cleared our own coast'. Yet, although in the end this proved a serviceable response, it was evolved in a context of muddle and bafflement. The most general expectation, contrary to the intelligence the English had at their disposal, was still that the Armada would make a direct attack on the south coast, probably at Portsmouth, Southampton, or, most probably, the Isle of Wight, 'the place which the enemies of this realm have principally desired'. Many of the leading soldiers, including

'The engraved image of the great-hearted Charles Howard', with allusions to the Armada and the Cadiz expedition of 1596 in the background. 'God gave the invincible fleet of Hispanic Philip vanquished into your worthy hand', reads the inscription on the scroll at bottom left. The subject is depicted in 'Imperial' style on a rearing war-horse (cf. p. 73). His signature appears below the engraving. Howard's career made him something of a popular Protestant hero: engravings like this were made to be sold 'at the Horse-shoe in Paternoster Row'.

The adoption of a squadron-formation by the English fighting ships is illustrated by this engraving by Pine. No source explains why the English resorted to this method of organization: it may have been in imitation of their enemies.

Sir John Norris and Sir Roger Williams, expected an attack on Plymouth 'for it is unlikely that the king of Spain will engage his fleet . . . before they have mastered some good harbours of which Plymouth is the nearest to Spain and easy to be won speedily, to be by them fortified and convenient to some succour either out of Spain or France'. By 1 August, Sir John Popham was convinced that no direct attack was intended, for the forces of the Armada 'of themselves are not fit, as I think, to enter England until they adjoin themselves to greater aids'. Yet Lord Henry Seymour, who commanded the straits fleet detailed to keep Parma in port, was equally convinced that no junction with Parma was intended. He seems to have felt that Parma's intense preparations were a feint designed to make the English split their force and so permit the Armada a victory; he was unsure where Parma was bound, but 'England I least doubt [i.e. *expect*]'.[10]

The Spaniards' real destination—the mouth of the Thames, as intelligence confirmed—was not neglected. Burghley claimed to be unable to sleep for worry about the defence of the Thames. £1,470 were spent on a boom across the river, which collapsed under its own weight. But even at the Thames mouth English predictions were awry. 'The greatest doubt', the deputy lieutenants of counties were told, 'is that the enemy will attempt to land in some part of Essex.' Hence the concentration of the main defence force at Tilbury. In fact, as we know, Parma intended to land on the south bank, apparently because he thought there would be superior foraging there and because the north bank was thought to be good cavalry country—a commodity which, for want of horses, Parma would have in short supply. His fears of the English cavalry were groundless: England had only a small and out-of-date 'feudal' cavalry supplied by gentlemen and their retainers. The English miscalculation, however, was, by far the more serious. A memorandum of the end of June 1588 warned the Council that Parma must be bound for Kent as 'his flat-bottomed boats are not to be adventured upon the seas, but in the shortest of passages, and in fair weather', but this expectation does not seem to have been widely shared. The bridge that was to have carried the English army to Kent at need was still unbuilt when the Armada sailed by; and even the army was still unassembled—or, at least, well short of the ten thousand men envisaged, when the campaign was effectively over.[11]

English strategy was hamstrung by ignorance of enemy intentions. The Spaniards, too, suffered from serious misconceptions about English preparedness. When Medina Sidonia warned the king, 'I felt that this undertaking was being represented to Your Majesty as being easier than it really was for certain ends and purposes not being in Your Majesty's true interests', it seems likely that he was referring to two sources of information which consistently misled the king: the English Catholic exiles, who were desperate to see their country saved and their fortunes restored and who were readily given, by virtue of their plight, to the easy luxury of wishful thinking; and Bernardino de Mendoza, Philip's ambassador in Paris, who allowed his hatred of England to occlude his

William Borough, clouded by his conduct at Cadiz in 1587 (see p. 96), was posted in the Thames with the queen's galley during the Armada campaign. He wrote to Walsyngham 'If it shall be thought good to appoint me to further charge, to command any of the ships that shall guard the river, what shall be directed me therein [I shall] observe and perform dutifully, with God's help, whilst life lasteth. I do send herewith a rough plot of this river's mouth, with the channels and shoals in it, to the end I may be appointed in the same plot where the ships should remain.' The dotted lines in this copy represent channels suitable for shipping. 'Your Honour may perceive', continued Borough, 'how they concur at the east end of the Nore.'

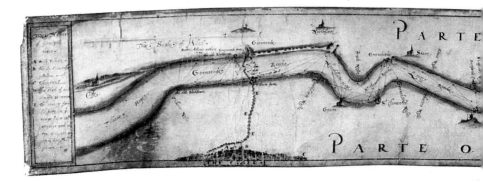

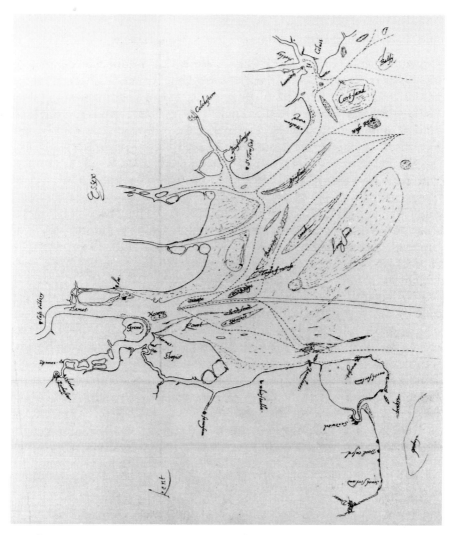

In Robert Adams's plan of the Thames defences, South is at the top: Gravesend, with the boom clearly depicted, is therefore on the left and London on the right. Note the sight-lines drawn from vantage-points on the bank.

judgement and blind his perception. He had a formidable network of informers in England, who, in the way of such men, fed him the information they knew he desired. And so dispatches and memoranda filled Philip's head with tales of leaky English ships, feeble English manpower, and rebellious English subjects. This misinformation, however, might have been harmless had it not been grafted on to a misconception. The Spaniards were well aware that English tactics were based on gunnery and that the English would not attempt to grapple and board. Yet all Spanish thinking about the likely course of any battle assumed that only grappling and boarding could be decisive. Knowing, therefore, that the Armada's contingent of fighting men would vastly outnumber anything the English could muster, Philip II was satisfied that his forces would be superior in this decisive respect. 'According to the reports I have here, which are correct,' he wrote to Medina Sidonia at the end of April, 'you will understand how superior are the forces you will carry to those which await you.' On 21 May he repeated, 'Their fleets are very inferior in forces, all told, to those which you carry.' On 1 July he advised, 'The ships of the enemy are all old or small, different from those we make here in quality and soundness'—here one can detect the influence of Mendoza's 'advices'—'leaving aside the advantage our men have in their numbers and in the experience many of them have.' All strategic thinking in Madrid was based on these erroneous assumptions.[12]

An almost imponderable element in the strategy of both sides was the likely behaviour of the Dutch. Potentially the Dutch occupied a position of commanding influence, for they had huge numbers of low-draughted fighting ships which posed a terrible threat to any sally Parma might attempt from the Scheldt. Intermittently, since 1585, they had kept up a fairly daunting patrol, effectively blockading the coast from Zealand to Calais since 1585; they normally had a dozen ships outside the Flemish ports, especially Dunkirk; after Sluys fell to Parma, two vessels always kept watch there; and five squadrons of five ships each plied back and forth towards the Channel. After the episode of the Armada was over, the Dutch claimed credit for England's safety:

For that our fleet, under the charge of Count Justinus of Nassau, being happily arrived and riding off Dunkirk at the very time of the rediscovery of the Armada of Spain, the forces of the Prince of Parma, then ready to put to sea, were by the same closely locked in and stayed within the said Dunkirk.

There are good reasons for believing this, even though Howard declared that at the material time 'there is not a Flushinger or Hollander at the seas'. In the first place, Dutch goodwill towards England, if not absolutely

reliable, was firmly grounded on common interest and effective alliance
in war against Spain. The Spaniards expected Anglo-Dutch co-operation:
indeed, they overestimated its extent, expecting to find Dutch contingents
guarding the Thames. When Lord Henry Seymour was ordered to the
Channel guard, his brief had been expressed as reinforcement of the Dutch:
the English expected to find some of their allies, at least, *in situ* when
Edward Bellingham was sent out with ten ships of the Merchant Adventurers to replace Seymour after the battle against the Armada: Bellingham
was 'to impeach the coming out of the Duke of Parma with his forces,
and to associate himself with such ships of Holland or Zealand or with
such other ships as the Lord Admiral hath left there'.[13]

On the other hand, Spanish plans seem to have taken remarkably little
account of the Dutch, perhaps not altogether in error. It was, in the first
place, doubtful how far the Dutch would expose their own forces in order
to save the English. The two peoples were uneasy allies and Philip II was
surely right to point out that 'the rebels in Holland and Zealand care more
for their own interests'. Elizabeth found there was 'no such ungrateful
people on earth'. She had threatened often before to abandon their cause,
and they were at liberty to do the same. For Parma to be embroiled in a
second front in England would not necessarily be to the disadvantage of
the Netherlandish states. English sovereignty might be the upshot of an
English victory; and when Elizabeth had reminded the Dutch that they
were 'simply ordinary persons in comparison with princes', she had left
them with the impression that it would be no improvement to them to
exchange English for Spanish rule. In any case, even had the Dutch put
their hearts into an attempt to frustrate Parma—and it must be said that
they are likely to have done so, since his bargeloads of soldiers would be
better, from a Dutch point of view, at the bottom of the sea than in
England—it is by no means certain that their intervention would have
made much difference. Parma had built up a strong force of forty flyboats
and armed pinnaces with which to break their blockade. He had constructed a canal system which enabled him to concentrate them unobserved; and he planned a feigned sortie from the Scheldt to free his intended
passage from Dunkirk. In December I 5 8 7, he had entertained hopes of
slipping by them unobserved: that was no longer possible in the conditions
of August I 5 8 8, but he might have been able to fight his way out. Drake
believed that, in the right conditions, Parma was capable of making the
crossing on his own. The examination of Don Diego Pimentel by his Dutch
captors reveals that Spain had been willing to contemplate huge losses, if
necessary, in running the gauntlet across the Channel. Parma himself, as
a good husbandman of his own troops, may not have been willing to

throw their lives away; but in his correspondence with the king, he never represented the Dutch as the major threat: it was Seymour's galleons that would certainly have been able to smash his flyboats and barges like matchwood.[14]

In any case, Dutch strategic planning seems to have been no more coherent than that of the English or Spanish. Like their allies and enemies, they were bewildered by the options and, even more than the main protagonists, divided in their high command. Their revolt against Spain had been started and sustained by a spirit of local particularism. Few provinces would sacrifice their autonomy to central planning and those which provided the fleet—Holland and Zealand—were the most jealous of all. They were also normally the least willing to co-operate with England, partly because it was ports of their own—Flushing and Brill—that the English occupied as surety for Dutch debts, partly because their merchants stood to profit from trade with the Spanish monarchy, and more generally because on many routes the English were their commercial rivals. Even if they took action against Parma they seem to have been determined not to do so along lines dictated by English interests. England's policy of blockading Parma seemed guaranteed to leave in the Low Countries a malignant cancer which the Dutch hoped to excise. The strategy favoured by the commander of the Dutch navy, Justin of Nassau, seems rather to have been to tempt Parma out into the open sea before acting. The arrival of the Armada in the Narrow Seas made this option seem less attractive. On 6 and 7 August Dutch ships emerged from their apparent lethargy in Flushing and cruised off the banks. Their appearance was both timid and intimidating: timid because they stayed inshore, out of the reach of Spanish galleons; intimidating because, had the weather favoured a sortie by Parma, they would have been on hand to attack him, if they chose. On balance, it is fair to say that the Dutch 'blockade'—such as it was—has been given too much importance by historians of the Armada. At best, it deserves third place among the reasons which stayed Parma's hand, after his own unpreparedness and the adverse weather.

The muddled evolution of Spanish strategy up to the time of the sailing of the Armada had settled on a 'junction' between Parma and the Armada as the central feature of the plan, but had left unresolved the question of how this was to be effected. The crucial problems were how was Parma's force to be protected during its crossing? And how and when was an anchorage for the Armada to be secured on the English side of the Channel?

A rational procedure would have been for the Armada to seize an English haven at an early stage of the campaign. Parma seems to have

At the time of the Armada, Justinus of Nassau, the Dutch naval commander, was a young man of twenty-nine. His mature portrait evinces gravity, sobriety, and sincerity. The simplicity of his lace collar should be compared with Parma's on p. 84. This work is conventionally attributed to the studio of Jan van Ravensteyn, but may belong to that of Michiel van Miereveld (1567–1641), whose 10,000 meticulous portraits included many of Justin's family.

On this panel painting the Spanish fleet is shown aflame, and the Dutch allies—who were good customers of the coastal Norfolk parish where this painting hung—are given a prominent role. The note of secular triumph which emanates from the central parade is mitigated by the religious texts with which it is embellished. In the upper panel the queen appears at prayer, and in the lower angels sustain a speech of the queen's which reviles the Spaniards as Philistines and praises her own 'Host of the Living God'. 'And if God do not charge England with the sins of England', the text continues, 'we shall not need to fear what Rome or Spain can do against us, with whom is but war and flesh, whereas with us is the Lord our God.' The frieze beyond the frame reads: 'The pope to God, Mendoza [perhaps an error for Medina Sidonia?] to Drake, and Philip to Eliza succumbed. England was victorious, Spain vanquished in the year 1588.'

assumed that this would happen, apparently incredulous that so great a fleet could operate so far from home without a safe harbour. 'I have always supposed,' he wrote, when the Armada was in the North Sea and the campaign effectively over, 'that the duke would have managed this as speedily as possible on his way up.' Parma was not simply being wise after the event. Without such a base, he told the king as early as September 1587, the Armada could achieve no more than a diversionary role. Yet from about that time, any thought of employing the Armada in this way seems to have been banished from Madrid. Memoirists who favoured a northern base for the Armada continued to issue warnings and advocate alternative plans. As late as April 1588, Bernardino de Escalante was urging a return to the plan of mounting a direct invasion of England from Spain on the grounds that the Flanders harbours offered no anchorage and were blockaded—or at least impeded—by the Dutch. That same month, Parma again asked to be allowed to resume his land operations against the Dutch rebels, so as to try to capture Walcheren, or at least Flushing, before the Armada sailed. These pleas seem to have been ignored. Plymouth and the Isle of Wight had been discussed as possible destinations; Plymouth, particularly, aroused English anxiety 'for it is unlikely', as Sir John Norris opined, 'that the king of Spain will engage his fleet ... before having mastered some good harbours, of which Plymouth is the nearest'. And even after the possibility had been excluded from Spanish strategic thinking, the captains of the Armada felt sorely tempted by the prospect of taking Plymouth as they approached it; only, it seems, by Medina Sidonia's decision to follow the course dictated from Madrid, did a council of war decide against it. The results were predictable: Sir William Wynter accurately foresaw that 'these huge ships that are in the Spanish army shall have but a bad place to rest in, if they come so low as to the eastward of Plymouth'. The Isle of Wight was thought by some Spanish pilotage experts to be likely to provide a suitable anchorage, though Parma seemed dismissive of it and it had a bad reputation with most pilots. According to the interrogation of Don Pedro de Valdés, it was to be used as an emergency anchorage in case of bad weather or need of repairs; it might be taken, the king allowed, after Parma had crossed, or if his crossing proved impossible, 'but on no account should you attempt to capture the Isle of Wight on your way eastwards without first having made a supreme effort to achieve success in the main task'. The unease this order inspired in the Armada's captains is apparent from their collective decision in a council of war off the Lizard on 30 July 1588, in the very mouth of the Channel, that they would try to avoid going beyond the Isle of Wight until they were sure of effecting a meeting with Parma, because of the danger to which lack of a safe harbour *en route* exposed them.[15]

Medina Sidonia, in particular, was evidently dismayed at his orders to operate without a base at hand. 'I greatly wish', he wrote to Parma on 10 June 1588, 'the coast were capable of sheltering so great a fleet as this, so that we might take some safe port to have at our backs; but as this is impossible it will be necessary to make the best use we can of what accommodation there is.' The inadvisability of this was demonstrated in the event, when the Armada was forced to extemporize an anchorage in the Calais roads on 6–7 August with the familiar disastrous results (below pp. 185–9). Yet it was a point on which the king's orders were implacably clear: Medina Sidonia was to allow nothing to deflect him from his course through the Channel. That this was later recognized as a fatal mistake is shown by the revised strategy adopted when Spain attempted a re-enactment of the Armada campaign in 1597: on that occasion, the new Armada was to secure Falmouth as a base before engaging in further operations.[16]

No Spanish document explains why the need of a haven was so sedulously ignored. Perhaps to try to seize a port was thought too great a risk. Perhaps the loss of time or the danger of alerting English defences was a decisive consideration. Perhaps Plymouth was thought too strong and the pilotage round the Isle of Wight too hazardous. Perhaps it was thought that the Armada's strength would be dissipated by the attempt. Whatever the reason, it was evidently felt in Madrid and the Escorial that a port 'at the Armada's back' was unnecessary. To understand this, it is necessary to broach the second great unanswered question of Spanish strategy: how was Parma's crossing to be protected?

In theory, there were four ways in which Parma might be given a clear passage: the English could be defeated, or diverted, or cowed; or the invasion force could be escorted in a form of convoy, as the bullion fleets were across the Atlantic, or as the Armada's own supply vessels were up the Channel. Diversionary strategy, as we have seen, had been ruled out in the late summer or early autumn of the previous year: it was recognized in Spain that the English would not be deceived by a feint but would continue to blockade Parma as long as they had shipping for the job. By far the best plan would have been to try to defeat the English fleet at sea before the proposed junction with Parma. So much, indeed, does this seem the obvious strategy to adopt that many historians have assumed that it was what the Spaniards had in mind. Yet the evidence suggests the very opposite: though there were circumstances in which the prospect of a battle was contemplated, the Spanish plan was to avoid it if possible. While the king's instructions were ambiguous, and Spanish policy-makers in general hesitant on this point, Medina Sidonia, at least, interpreted his instructions to mean that he should not give battle, save in necessity or at an overwhelming advantage. This impression was consistent with most of the instructions issued by the king since the time of Santa Cruz's command. The marquess had been given permission to give battle if the English opposed his course up the Channel, or if he encountered a detachment at the mouth of the Channel, near Ushant or the Scillies: in those circumstances, 'if they are divided, it will be well to proceed to overcome them so that they cannot all join forces'. Such permission was, however, expressed conditionally and even the attack on a divided English fleet was limited to the most westerly portion of the Channel. The instructions are quite explicit, however, in stating that battle is to be avoided if Parma can be got across without one: 'this point about fighting is to be understood to mean if a crossing to England cannot be assured for the Duke of Parma by any other means, for if this passage can be granted to Parma without fighting, by drawing the enemy off or by some other means, it will be well'. At this point in the document, an annotation by the king reads: 'this is well, although I doubt whether it can be done without fighting, but let it stand as it is'. These instructions were passed on almost word for word to Medina Sidonia on 1 April 1588: 'if Drake should pursue and overtake you, you may attack him, as you should also do if you meet Drake with his fleet at the entrance to the Channel, because if the enemy's forces are so divided, it would be well to defeat him by stages and thus prevent those forces from uniting'. And again: 'It must be understood that the above instructions about fighting only hold good in case the passage across to England of my nephew the Duke of Parma cannot be otherwise assured. If this can be done without fighting, it will be best to carry it out in this way, and keep your forces intact.' When the king wrote of the importance of 'striking at the root' he apparently meant not destroying the English fleet, but landing the army on English soil. Medina Sidonia, following the king's lead, seems to have been equally unwilling to see these two goals as interdependent. All the king's communications with him stressed the need to proceed directly to 'join hands' with Parma; none highlighted a battle other than as a hazard to be avoided or a necessity to be regretted. Medina Sidonia sailed in the firm conviction that he was not expected to fight the English until after he had joined Parma. Juan Martínez de Recalde—and doubtless many other professional naval commanders as well—sailed with exactly the opposite conviction; 'So far as I understand,' he wrote, 'the object of the Armada is to meet and vanquish the enemy by main force, which I hope to God we shall do, if he will fight us, as doubtless he will.' Only if the enemy refused battle were they to proceed directly to the hoped-for 'junction'. But Recalde's was the voice of experience and common sense. He had not seen the plan drawn up in Madrid.[17]

As it turned out, there were in any case few opportunities for the Armada to fight a decisive battle on its way to the illusory rendezvous. The best chance the Spaniards had to settle the contest by battle occurred on 30 July, the day after entering the Channel, when the Armada was approaching Plymouth. At that time the Armada had the initiative in three crucial respects: the Spaniards had the wind behind them; the enemy were unaware of their position; and a portion of the English fleet which faced them, heavily outnumbered by every measure, was confined to Plymouth harbour or Sound by the adverse wind. Had an early battle been part of the Spanish plan, this would have been the moment for it. It required the exercise of only a modest initiative beyond the limits specified in the king's orders, which had anticipated the possibility of an irresistible opportunity for a battle at overwhelming advantage; the king had, moreover, stressed the usefulness of defeating the enemy piecemeal, and had specifically allowed for a battle should the English be met off the Lizard—where the Armada had been only a day before. But the general hostility of Medina Sidonia's brief to the idea of a battle, and the specific order to engage the enemy only if he barred the way head-on, seems to have deterred the duke from seizing his chance. It was a near-run thing. According to the testimony of Don Pedro de Valdés, when a captured fisherman revealed that the English were in Plymouth, 'the duke called a council to consider of entering there and conquering the said fleet'. The two most experienced naval commanders—Oquendo, impetuous as ever, and Recalde—both favoured an immediate attack, as did the most esteemed land general, Alonso de Leiva. De Leiva's plan may have involved landing an infantry force to make a concerted amphibious attack on the town. Other voices were raised against the idea. Don Pedro himself 'was of opinion that it was not fitting to do so, because that the fleet was within the haven, whereof the mouth is so strait as not more than two or three ships would go in abreast, which was insufficient for that action'. Medina Sidonia was opposed to an attack, partly because he felt it was against the king's orders and partly because he feared his fleet would be exposed to strong shore batteries if it tried to sail through the neck of Plymouth Sound. The Spaniards' failure to act was doubtless also the result of understandable misgivings about the reliability of the intelligence they were debating. They seem to have taken no positive decision to renounce an attack, but rather simply to have temporized, continuing their approach while postponing decisive action. In a later account to the king,

Valdés claimed that 'it was resolved we should make to the mouth of the haven and set upon the enemy if it might be done with any advantage; or otherwise keep our course directly to Dunkirk without losing any time'. More quickly than they can have believed possible, the Spaniards' chance vanished. Within two watches of the breakup of their council of war, the English fleet suddenly appeared in their rear, having stolen the weather-gauge. The English had warped out of Plymouth, straining at their oars, and tacked against the wind until they were behind the Armada. The manœuvre had been executed overnight with daring and speed, as soon as the adverse tide-race slackened. There can only just have been time for Drake's mythical game of bowls to be finished with the leisure alleged in the popular story.[18]

The opportunity that had come and gone when the English were penned in Plymouth was never to be repeated. On the open sea, the English could always exploit the manœuvrability of their ships to elude battle at will. Thus though Medina Sidonia tried to take the next two chances which offered—once when the wind veered and the Spaniards briefly had the advantage of the weather, and once in a dead calm when the duke sent in the galleasses—he was never again in full possession of the initiative

The approach of the Armada, not yet drawn up in battle order, towards the Lizard on 29 July, in Robert Adams's version. Notice the Spanish pinnace reconnoitring the coast and the English vessel out of Plymouth spying on the Armada.

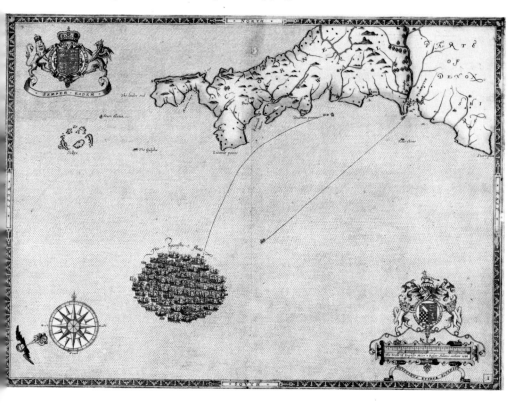

and never able to dictate the terms of battle. A battle in Plymouth Sound would have been costly, but it might have been decisive. Without room to escape, some English ships could have been grappled and boarded and the superior numbers of the Spanish soldiery would then have told.

Given that neither to divert nor defeat the English were essential parts of the Spanish strategy, what alternatives can the Spanish decision-makers have had in mind? On the whole, it seems that the king and his advisers hoped the English would be intimidated into withdrawing without a fight, while Medina Sidonia (who, if he ever believed this, ceased to do so once he was in the Channel and had witnessed English methods) seems to have hoped to escort Parma across the Channel in convoy in the teeth of the English fleet. Both expectations were equally fantastic. The king's instructions repeatedly assigned a meeting-place 'at the Cape of Margate'—that is, the North Foreland—off the Kentish coast at the mouth of the Thames. On the face of it, this seems a manifestly foolish notion, incompatible with the fact that Parma was unable to make a sortie from Dunkirk. Yet the implication is unmistakable in the king's messages that by virtue of the very presence of the Armada off Margate, Parma's passage would be free. As early as 4 September 1587, Philip told Parma that the Armada would be able to 'secure' his passage from a position at the mouth of the Thames. This expectation was echoed in the crucial passage of Medina Sidonia's instructions, which stressed that no change of plan was to be made from what had been communicated to Santa Cruz in September: 'You will sail with the whole of the Armada, and go straight to the English Channel, which you will ascend as far as Cape Margate, where you will join hands with the Duke of Parma, my nephew, and hold the passage for his crossing'. The king acknowledged that a battle might have occurred; but was hoping it would be unnecessary; he continued to hope that the enemy might be caught 'unprepared', though this must have seemed unlikely. The best hope of an assured passage lay, as the king and his advisers thought, in the timorous withdrawal of the English.[19]

It must be remembered that Philip was convinced of the superior battle-worthiness of his own fleet and in particular, though he realized that English battle tactics were based on fire-power, not manpower, he continued to think that the Armada's superior numbers of fighting men would be a decisive advantage; the English would therefore avoid battle altogether or defer it until after the invasion force had landed. These predictions were made, with every sign of confidence, in Medina Sidonia's supplementary instructions of 21 May 1588, though the king added, with scarcely more realism, 'If in spite of this they are ready to fight, you have orders in the instructions to attack and attempt to destroy them, in which

God will assist you.' Most documents, however, deferred this contingency to a remote area. Filippo Pigafetta, the strategic expert called in by the king to advise on the Armada, assumed the English would not dare to fight while the full Spanish complement of soldiers was aboard the Armada. The king was confident that once his fleet had gained the mouth of the Thames, 'and is thus able to prevent the union of the vessels in the river of London and on the east coast of the island with those on the south and west coasts, the enemy will hardly be able to collect a fleet with which he could dare to seek out ours'. Thus Parma would 'see his crossing made secure by the presence of the Armada at Margate'. This plan was obvious landlubber's nonsense: the Armada would not have been able to prevent the reunion of the detachments of the English fleet by anchoring off the Thames estuary; and Parma's force would have been vulnerable even to a tiny fragment of Elizabeth's navy. But it seems that some at least of the Spanish planners, and probably the king in particular, were, for a while and at a material time, persuaded by it. It seems to be assumed in the king's letters to Medina Sidonia of 21 May: 'Their fleets are very inferior to yours ...' the English may defer their attack until after the invasion 'with the aim and hope that when you have disembarked men on land your Armada will be weak ... What you have to do is follow your course as you have been ordered straight until your junction with the duke my cousin and join both forces together without deviating previously on any account, other than to smother and destroy whatever may obstruct your path.' The 'day of battle'—the king is unequivocal on this point—must be ventured before the junction with Parma is effected only if 'all the forces of the enemy come in front of you'. Otherwise, such an encounter was to be avoided until after the invasion. Even Parma, who seems to have counted on the Armada to clear his passage by battle, occasionally succumbed to unrealistic expectations of the fleet's intimidating effects. In a letter to the king of 7 August 1588, when the Armada was in Calais, he professed to believe, although 'the enemy has a large force of armed vessels on this coast to oppose our coming out', that 'doubtless they will depart when the Armada arrives'.[20]

Failing the destruction, diversion, or withdrawal of the English ships, the last means by which Parma's crossing might in theory have been secured was a convoy. There is no trace of such an idea in the early stages of Spanish planning; but just before the departure of the Armada, and during its course, two events occurred to make such a manœuvre seem possible. First, Parma announced the completion of his canal system, adding that this made it unnecessary for the Armada to proceed 'so far up the Channel': he could now easily move his Sluys force in security and safety to Dunkirk. This may have been interpreted in Madrid, and by Medina Sidonia, to mean that the rendezvous, formerly fixed for off the North Foreland, could be anticipated. Filippo Pigafetta, for instance, argued that a fleet could lie in summer between Sandwich and Dover, where Caesar disembarked. Henceforth, the 'Cape of Margate' ceases to be mentioned and it seems fair to suppose that the exact whereabouts of the proposed 'junction' was now to be left to the discretion of the commanders on the spot. Recalde and Don Pedro Valdés both spoke throughout the voyage of heading towards a position as near as possible to Dunkirk, which indeed seems to have been Medina Sidonia's trajectory. Spanish prisoners later assured Drake that Dunkirk was the 'place appointed' for the meeting of the Spanish forces. Secondly, the experience of the voyage up the Channel showed that the Armada was impregnable to English attack, so long as the fleet's formation was preserved. As Howard had admitted, the English 'durst not put in among them'; and no Spanish ship was lost to enemy action during the long pursuit of the Armada between Plymouth and Calais, despite the expenditure of a 'terrible value of great shot'. During that period, the Armada had effectively acted as the convoy of its own supply ships, hedging them about with its galleons, fighting off English attacks, and successfully sparing stragglers from being 'cut out' by the assailants. The system had worked—and would continue to work—on the Atlantic bullion fleets, against which only three successful raids were made in a century and a half of navigation. It might have been possible to envelop Parma's troop-carriers in the same protective covering.[21]

The idea of a convoy had, at any rate, more chance of success than the plan of defying the English from the North Foreland in the hope that they would be too timid to obstruct Parma's passage. It had, however, two fatal flaws: first, Parma seems to have been reluctant to co-operate in it; secondly, and more fundamentally, because it was extemporized at a late stage, and had played no part in the planning, the Spaniards found they lacked suitable shipping for it. Medina Sidonia may have had a notion of something like a convoy in his mind even before he set sail. We know, at least, that he had the inkling of an idea that might have contributed to such a notion: he believed that Parma could sail out of Dunkirk to meet him in the Channel, and be escorted on from there. This suggestion was rebutted in indignant letters of the Duke of Parma, protesting that such a manœuvre would be impossible:

The duke ... seems to have persuaded himself that I may be able to go out and meet him with these boats. These things cannot be and in the interests of Your

Majesty's service I should be very anxious if I thought the duke were depending upon them ... With regard to my going out to join him he will plainly see that with the little, low, flat boats, built for these rivers and not for the sea, I cannot diverge from the direct passage across which has been agreed upon. It will be a great mercy of God, indeed, even when our passage is protected and the Channel free from the enemy's vessels, we are able to reach land in these boats. As for running the risk of losing them by departing from the course agreed upon, and thus jeopardising the whole undertaking, if I were to attempt such a thing by going out to meet the duke, and we came across any of the armed English or rebel ships, they could destroy us with the greatest of ease. This must be obvious and neither the valour of our men nor any other human effort could save us.

   With prophetic misgivings, Philip wrote in the margin of this letter, 'God grant that no embarrassment may come from this.' Parma referred to letters in which he acquainted Medina Sidonia with his objections. The duke seems never to have received them. Nor, in this instance, was the substance of what Parma said passed on to him by the king. As he sailed up the Channel he formulated his plan for a joint crossing with Parma in convoy in fatal ignorance of Parma's case against it. Medina Sidonia, quite against his character, and quite beyond the normal limits of his intellect, broke the stagnation of Spanish strategy with a plan which, in other circumstances and with other ships, might have offered a prospect of success. But this last plan was as flawed with impracticable elements, and as bedevilled by misunderstanding between the individuals involved, as all those which preceded it.[22]
   There was one device which might have overcome the technical problems that made a convoy impossible. The Armada was unable to approach the ports Parma held because deep-draughted ships could not navigate the Flemish shoals; Medina Sidonia could therefore only hope for Parma to 'come out to meet him', which in turn, because of the blockade and the general fragility of his troop-carriers, he was unable to do. The twain could not meet. What sort of vessel could have traversed that gap? What sort of ship combined the fighting power to drive off enemy flyboats, with a depth of draught that could be accommodated close inshore on the Flemish coast? The obvious answer would have been the galley. Galleys had been used by the Venetians and Florentines in their Flanders trades for centuries. A few years later, the campaign of Federico Spinola in the Channel and North Sea was to show not only that they were serviceable craft for northern warfare, but also, specifically, that they could beat flyboats in almost any conditions, and even overcome galleons if the conditions were right. Furthermore, Spinola showed that the port of Dunkirk—the crucial place for the embarkation of Parma's army—could

be readily modified to handle galleys. They were the sort of ships that could be used either as troop-carriers, if available in sufficient numbers (65 could have carried Parma's entire force), or to tow Parma's barges; they could have helped his fleet warp out of its Flemish bases in a calm. And to judge from Spinola's experience, they might not have been entirely useless had the Armada engaged the English on the open sea, for in favourable conditions they could match English speed and manœuvrability. They could do twelve knots in a dead calm. A Spanish spy—admittedly, an unreliable one—even claimed of the English that 'the only fear these people have is that they should be attacked by galleys'.[23]

Yet the Armada had no galleys. The original plan of the Marquess of Santa Cruz had called for forty or fifty of them. In March 1587, Medina Sidonia had recommended that the fleet should have eight galleys at least 'since to disembark troops on land and to tow the barges that carry them for this purpose they are of the utmost necessity'. When he knew that he would have to command the fleet, his pleas for galleys became urgent. 'It is of the utmost importance that galleys should go with the Armada', he told Juan de Idiáquez, 'and it will be well, as you say, to take four of the Spanish galleys for that purpose, or even eight, which, joined with those at Lisbon, would be twelve. They would be of the greatest use and value.' His recommendation, however, was ignored. Only the four galleys of the Lisbon squadron sailed with the Armada and they all retired from the fleet before reaching the Channel. Historians have generally considered this as confirming the uselessness of galleys, and it is true that the Lisbon squadron seems to have been all too easily deterred by the unseasonably bad weather of the summer of 1588. But with vessels of Italian build, used to the northern run, and crews well seasoned and selected, galleys could have performed creditably and perhaps decisively, as in the Spanish raid against Cornwall in 1595 and Spinola's campaign from 1599 to 1603. Galled by Spinola's success, the Dutch and English began to build galleys in reply; inspired by his example, Monson was to recommend that England revive her galley force in the 1630s. Unfortunately, at the time of the Armada, galleys were momentarily unfashionable for northern warfare. Perhaps the main reason for this had been their failure in battle against Drake's galleons in the bay of Cadiz in 1587; but on that occasion the galleys had been hopelessly outnumbered and unsupported by larger craft; the Cadiz experience was wrongly interpreted if it was understood to mean that galleys could not make a useful contribution as part of a combined fleet. Yet Fenner's view that 'twelve of Her Majesty's ships' will 'make account' of 150 of them seems to have been widely shared. Howard thought the English navy's sole galley was only 'fit for the fire'. Historians

have been infected by the same contempt. Yet galleys might have won
the war of 1588 for Spain.[24]

The frustration of Medina Sidonia's hopes of a convoy can be traced in
his and Parma's letters. The king had never ceased to stress the importance
of close and frequent communication between the two commanders; yet
this was a necessity easier to prescribe than to supply. As Medina Sidonia
entered the Channel, he complained of Parma that 'I am astonished to
have had no news of him for so long. During the whole course of our
voyage we have not fallen in with a single vessel or man from whom we
could obtain any information, and we are consequently groping in the
dark.' A week later, with the Armada in the Calais roads, and Medina
Sidonia frantic for a rendezvous, the situation was no better: 'not only
have I received no reply to my letters, but no acknowledgement of their
receipt has reached me'. Yet now the Spanish forces were separated by
only a few hours—though Parma made things worse by lingering in his
camp at Bruges until the 8 August—and messages at last began to be
freely exchanged; they revealed, however, only an utter lack of mutual
comprehension. 'If you cannot at once bring out all your fleet,' wrote
Medina Sidonia to Parma on the 6th, 'send me at once the forty or fifty
flyboats I asked for yesterday, as, with this aid, I should be able to resist
the enemy's fleet until Your Excellency can come out with the rest, and
we can go together and take some port where this Armada may enter in
safety.' The gist of Parma's reply may be gleaned from a letter to the king:
'If the duke succeeds in getting to a place where I can help him, Your
Majesty may be sure that I shall do so ... To judge from what the duke
says, it would appear that he still expects me to come out and join him
with our boats, but it must be perfectly clear that this is not feasible.' Yet
Medina Sidonia was writing to Parma on that same day, 7 August, to
protest that it was very inadvisable 'for the Armada to go beyond Calais.
I beg you to hasten your coming out before the spring tides end, as it will
be impossible for you to get out of Dunkirk and the neighbouring ports
during the neap-tides.' This time, an annotation summarizes Parma's
reply: 'he may be informed that ... in certain states of the wind the water
goes down and the spring tide is necessary, but there are only a very few
boats which run this risk ... There has never been the slightest question
or idea of waiting for the spring tides, or of deferring the enterprise on
this account.' The following day, a letter of Parma's to the king shows
that he had still not conveyed his views successfully to Medina Sidonia:

I have news that the duke, with the Armada, has arrived in Calais roads [*in fact
Parma had known of this by the previous day*]. God be praised for this! Although it

may seem superfluous to insist upon a point which I know Your Majesty well understands, I cannot refrain from repeating once more what I have said so often already ... It appears that he still wishes me to go out and join him with these boats of ours, and for us, together, to attack the enemy's fleet. But it is obviously impossible to hope to put to sea in our boats without incurring great danger of losing our army. If the duke were fully informed on the matter, he would be of the same opinion, and would busy himself in carrying out Your Majesty's orders.[25]

It must be said that Parma, on his part, made little effort to understand Medina Sidonia's messages. Sometimes, indeed, he wilfully misconstrued them. He read the letters he received on 10 August, for instance, as primarily about 'the danger the Armada was in for want of gaining a port', although they were really about the execution of the plan as a whole, and seized the opportunity for inserting an unjust jibe: 'I have always supposed that the duke would have managed this as speedily as possible on his way up.' He also claimed to believe that Medina Sidonia wanted him to use his transports as warships 'which shows how badly informed the duke must be as to the character of our small, weak boats', though this was a travesty of the plain and obvious meaning of the messages from the Armada. On the other hand, this evidence has to be treated with caution, as showing neither folly nor malevolence on Parma's part, but simply prudence. He made these remarks in writing to his master, the king, and was indemnifying himself against future recriminations in the event of failure. Communications between the forces that were intended to collaborate in the Armada campaign were undeniably bad, but not necessarily irremediable, nor solely responsible for the débâcle.[26]

The problems, which were insuperable anyway, were compounded by the unreadiness of Parma's forces. Parma had worked hard and—as far as can be judged—with no niggardly will to prepare for the invasion; he had painstakingly gathered his flyboats and pinnace-warships by building and hiring, including sixteen hired from France and thirteen from Hamburg. He had built 130 barges; he had constructed the all-important canal system; it was impossible to keep the army up to strength for the length of time demanded by circumstances, but he probably had over 16,000 effectives on hand which, despite his assertion that '50,000 would not be too many for the enterprise of England', were probably enough for the job when supplemented by the reinforcements the Armada brought. It cannot therefore be claimed that he had stinted his preparations. It remains true, on the other hand, that at the critical time he moved with a deliberation that baffled Medina Sidonia's entourage. When Don Jorge Manrique arrived in Dunkirk to observe Parma's efforts on Medina Sidonia's behalf, he was astonished to find the embarkation had not

In this fine propaganda-piece produced for Parma by van Veen, the duke is portrayed as the new Hercules, with club and lion's pelt. Catholic Faith, with her cross and rosary, points him towards a hilltop shrine, more evocative of classical paganism than of Christianity because it represents not the Church but Fame, a point emphasized by the pantheon in the background. Unfaith and Heresy cower, mutilated and petrified, in the foreground. The scrolls pinned to the tree show Parma's Dutch campaigns: a pontoon bridge is clearly visible on one of them.

started and considered it would be impossible to accomplish it in time. Medina Sidonia's military adviser, Don Francisco de Bobadilla, was scandalized. 'If Parma had been in Dunkirk as you told me, with his eyes open to leave as soon as he saw our fleet, it would have been of some use, but otherwise impossible.' Bobadilla acknowledged that the Armada lacked the right kind of shipping for the shoal-bound shores, but put the blame firmly on Parma. If Parma had set sail as the Armada arrived in Calais, he alleged, 'we should have been successful'. Apparently, to the men in Medina Sidonia's entourage, Calais was so near Dunkirk that Parma should have been able to attempt a sortie: the deviation from the 'direct crossing' on which Parma insisted was small; the risk from Dutch intervention should have been acceptable. Even if he lost half his men, Don Diego Pimentel later told Dutch interrogators, Parma must surely at least try to emerge. When he failed to do so, the men of the Armada, and even some of those in the Army of Flanders, were left with a sense of betrayal. Waiting despairingly to embark with the troops, Don Juan Manrique vented his frustration in a bitter letter to Juan de Idiáquez: 'I cannot refrain from saying how the happiest expedition the world has seen has been defeated. The day on which we came to embark we found the vessels still unfinished, not a pound of cannon on board and nothing to eat.' He exonerated Parma, whose indefatigable efforts he acknowledged handsomely, reserving the blame for the bloody-mindedness of the Flemish workers. Others were less charitable.[27]

Parma's own devoted admirer, the memoirist Alonso Vázquez, who fought as a captain in Flanders under his hero's command, and wrote a partial but never untruthful account of the deeds of this modern 'Alexander', was himself at a loss to apologize for his commander. Parma's defence, he tells us, 'was that in twenty-four hours, with much to spare, the provisions and other equipment could be embarked, and the army no less; and although it is usual for there to be confusion at an embarkation, it has already been seen how quickly the seamen are accustomed to do these tasks'. Parma's ship's forecastle was not yet completed, but it was sufficient for its job; the really grave problem, Vázquez admitted, was that the cavalry and artillery were not yet in a position to embark. 'This was negligence . . . for which I know not whom to blame, because whoever saw the care and concern of Alexander could not believe that the fault can have lain in him.' According to Vázquez, the soldiers of the Army of Flanders were eager to embark: they wanted to 'measure pikes' with the English. 'But to my mind, for I was present there, it was a big risk to confide so powerful an army to vessels so weak and fragile as those which were provided, although everybody thought them adequate for transport but not for fighting.' Perhaps for this reason, Vázquez suggests, Parma called the embarkation off. 'His good fortune, as some men said, began to decline that day. When he suspended the operation, the soldiers' tongues began to murmur against him ... a very new experience for him.'[28]

The real reason for calling off the embarkation was avowed frankly—albeit with much special pleading—by Parma himself. In a long, exculpatory letter to the king on 10 August, he claimed that the embarkation had made great progress on 8 August and was proceeding on the 9th, when Jorge Manrique arrived with Medina Sidonia's letters. Parma then convened a council 'of practical sailors', apparently to convince Manrique that the sailing conditions made an immediate sortie impossible. 'While this was under discussion, the embarkation continuing actively the while, the prince of Ascoli arrived' with the news of the dispersal of the Armada by fire-ships and its course away from Calais to the north. 'What happened subsequently with the other ships is unknown, except that the English continue to follow them in very swift vessels, manned by good and experienced sailors.' The bear, in Drake's words, was 'robbed of her whelps'. Yet Parma was able to view his frustration with feigned resignation and ill-disguised relief. No one could be more grieved than himself, he wrote in elaborate consolation to his master, at an outcome that 'must come from the hand of the Lord'. Philip could now only look forward to 'the full fruition of your desires in good time'. He should be careful of his health. 'The great army, moreover, which Your Majesty has intact, should with God's blessing banish all cause for fear, especially as it may be hoped that by His divine mercy the duke and the mass of the Armada may not have suffered any further loss beyond that which I have stated.'[29]

Suspicion inevitably surrounded Parma. It was such a rare thing for so great a prince to be entirely loyal to his sovereign. He was suspected in Madrid of harbouring designs upon the crown of Portugal, to which his son, Ranuccio, had a highly presentable claim; Walsingham made him a secret offer of the sovereignty of the Netherlands; according to the examination of an Armada prisoner, Vicente Álvarez, 'it was a question among them [*aboard the fleet*] that if the Duke of Parma should conquer this land [*England*], who should then enjoy it, either the king or the duke? And it was suspected that it would breed a new war between them.' The States of Zealand had made much the same prediction to the queen. Yet there is no creditable evidence to support rumours of treachery by Parma. There was nothing sinister about his tardiness in support of the Armada. As we have seen, he was tepid about it from the first. He frankly preferred to pursue the reconquest of the Netherlands; he had never believed in the feasibility of the Armada project and must have been taken by surprise to see it get so far. Like Medina Sidonia, he had urged the king to desist; he had recommended making peace while there was still a chance; and he had spared no opportunity to disembarrass himself of responsibility in advance. 'I will not be held responsible', he wrote in so many words on 12 September 1587, 'for any checks that may arise', though he was willing 'to sacrifice myself for your service, as I have always done, hoping that Your Majesty's Christian merits and good fortune will be present to aid us'. On 5 April 1588, he warned, 'If the co-operation of the Duke of Medina Sidonia should fail me, both before and during my embarkation, as well as after, there will be no question of my succeeding as I desire to do for your service.' These expressions, and others like them were intended to indemnify him against an outcome of which he was never in doubt. 'I am only a man,' he had felt obliged to remind his king at an early stage of the planning, 'and I cannot work miracles.' What was remarkable in him was not any shortfall of effort, but rather the scale on which he was prepared to strive for an enterprise which 'must depend on the holy and mighty will of God, for the zeal and industry of men cannot suffice'.[30]

Six

The Days of Sparring

'You don't suppose the Spaniards, the greatest nation on earth at the time, would have started off a thing like that Armada without seeing that the captains of the ships were sensible men. Of course they wouldn't'. Once again, the confidence of George Birmingham's hero was sadly misplaced; or, at least, his facts were accurate but the inference one might expect to draw from them is false. However sensible the captains, the tactics at their command were ineffective. And that applies as much to the English as to the Spanish side. Neither contender had a tactical system that was capable of defeating, or even seriously damaging, the enemy. When the campaign was over, this was frankly acknowledged in England and Spain alike. The English, while rejoicing in their deliverance, did not spare themselves anxiety over the limits of their success or enquiries into the 'little harm' done to their enemies. In Spain, the sense of defeat was embittered by recriminations. Don Francisco de Bobadilla might have spoken for both sides when he said, 'One must have seen with one's eyes and touched with one's hands what has happened, in order to realise the mistakes that have been made in this undertaking. Everybody is saying now, "I told you so; I guessed as much." When the hare escapes, everyone gives good advice.'[1]

The Armada fight has often been represented as marking a new departure in naval battle tactics: an encounter which, because of its unprecedented scale, imposed experimental conditions on the contenders and caused new solutions to be improvised. In fact, however, Spanish tactics were borrowed without serious modification from age-old tradition, and those of the English had been evolved over a generation of experience of the use of fire-power in Atlantic warfare. Both sides' tactics were predictable and those of each were known in advance to the other. Of the supposedly unique feature of the Armada campaign—a gunnery duel between large ocean-going fleets—both sides had a recent example before their eyes in the shape of the battle of Terceira, at which the Marquess of Santa Cruz had dispersed a French fleet (containing eleven English ships), to make possible the Spanish conquest of the Azores. In Philip II's case, the example was literally before his eyes, for he had it painted on a wall

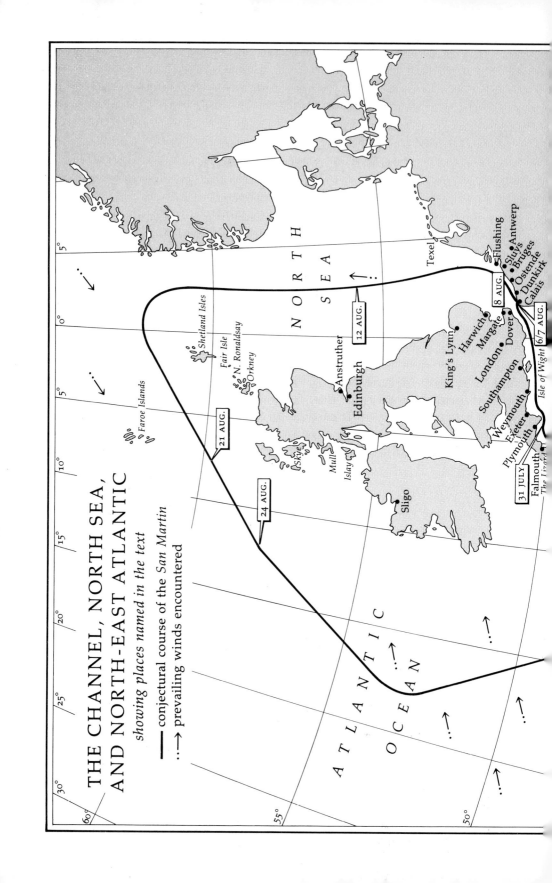

THE CHANNEL, NORTH SEA,
AND NORTH-EAST ATLANTIC

showing places named in the text

—— conjectural course of the *San Martin*

⇢ prevailing winds encountered

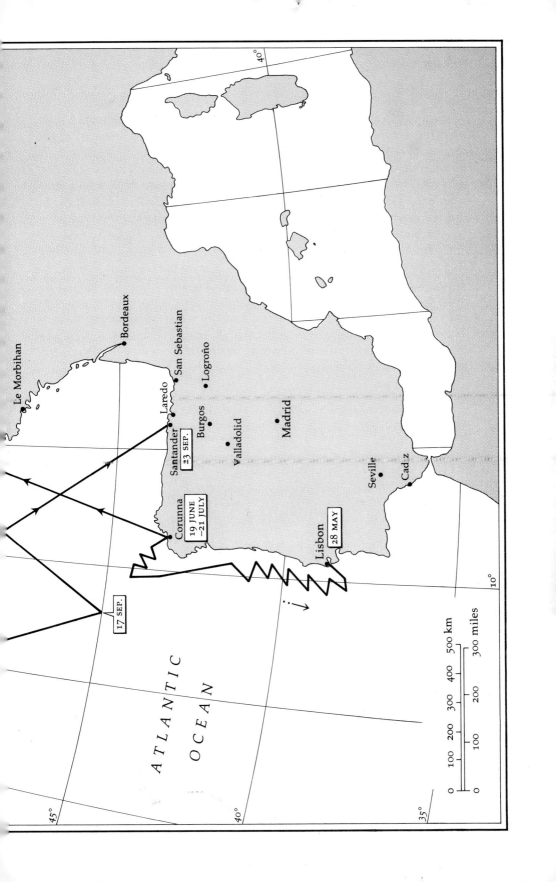

inside the Escorial. There he could see depicted the exchanges of heavy gunfire at close range, the havoc wrought by incendiary devices, the use of ordnance to sink ships; and as well as these things, expected and feared, he could behold an episode he must have hoped to see repeated in the case of the Armada: the troops rowing ashore in their landing craft, forming up on the beach-head, and scattering the enemy into the hinterland. It is rarely appreciated how closely the battle of Terceira foreshadowed the events of the Armada fight. The Marquess of Santa Cruz opened the decisive phase of the battle by gaining the weather-gauge by stealth, just as the English were to do on the first day of their encounter with the Armada. The scale of the battle, in terms of the numbers of first-rate ships directly engaged, was probably somewhat bigger in the case of Terceira than in any of the Armada battles; and while the Spaniards' main aim was to grapple and board enemy vessels, ship-smashing gunfire played a major role. The Spaniards actually sank more enemy ships at Terceira than the English did in the whole Armada campaign. When the French flagship surrendered, she was so low in the water that she foundered on the way to Spain.

Painted in fresco on the walls of his apartments in the Escorial, the scene of the battle of Terceira served to encourage Philip II's hopes and remind him of victory won by his fleet in a gunnery battle in which English ships took part. Note how ships are being disabled and even sunk by close-range gunfire.

Spanish landing craft disgorging their troops, *tercios* drawing up in phalanx in the face of a retreating enemy engaged by skirmishers: this scene from the conquest of the Azores, painted in fresco on the walls of the Hall of Battles in the Escorial, must have been much as Philip II would have imagined the culmination of the voyage of the Armada.

The experience of Terceira was, in one sense, delusive. It made Santa Cruz believe he could do the same to Drake as he had done to the French king's admiral, Strozzi. Other evidence, however, was also adduced in Spain and enabled the Spaniards to foresee with astonishing accuracy exactly what the English would do. Nearly a year before he took command of the Armada, Medina Sidonia warned the king of what would happen. Drawn from the recent experience of Hawkins's intended treatment of the Indies fleet in 1586, his letter could, after the first few lines, serve as a passable analysis of how the English treated the Armada. 'I recall', he wrote,

how John Hawkins, with six galleons of about four hundred tons and six other small ships, set himself to await the fleet which was coming from New Spain at the time, and how his aim was to gain the weather-gauge, which he would be able to do easily, as his ships were very light and without cargo, and he was sailing them newly clean and well rigged with new sails ... and once he had gained the wind, with the artillery he carried, very heavy-shotted and very much to the purpose for what he had in mind, to do us all the damage he could and to try to sink our ships of war and spoil the rigging of the rest, in order to make himself master of them, without their being able to resist and without being able to suffer any harm himself; for our galleons would not have been able to come to handstroke if the enemy did not wish, and he would have refused to do so.

All the essential elements of the English tactical system, as practised in the Armada fight, are here: gaining the wind, which was always a pre-condition of victory in the age of sail; laying off to use fire-power decisively; aiming to sink ships, where necessary, even at the sacrifice of some potential prizes; seeking a range that favoured English guns; relying on heavy ship-smashing artillery; avoiding—above all—grappling, boarding, and hand-to-hand mêlées; and using the swiftness and lightness of English ships, qualities for which they were chiefly designed, to do so. In 1586, Hawkins failed to strike, but there was no doubt on either side of what his tactics would have been had he waylaid the fleet.[2]

The Spaniards were also aware that the English would aim to fire low in an attempt to damage the hulls of enemy ships. If the report of the Portuguese spy, Fogaza, as early as 1574, had not sufficed, experience, by the time of the Armada, would have confirmed it. Philip II added warnings to the Armada commanders' instructions, in his own hand, about 'the way they have of firing low, as these reports inform us'. The instructions to Medina Sidonia—albeit supererogatory, since the duke had already demonstrated his knowledge of English tactics—summarizes the information at Spain's disposal:

You should take special note ... that the enemy's aim will be to fight from a distance, since he has the advantage of superior artillery and the large number of incendiary devices with which he will come provided; while ours must be to attack, and come to grips with the enemy at close quarters; and to succeed in doing this you will receive a detailed report of the way in which the enemy arranges his artillery so as to be able to aim his broadsides low in the hull and so sink his opponents' ships.

There can be no doubt that the implications of the English tactics were fully understood by the Armada's commanders. Juan Martínez de Recalde explained the prospects to the pope's representative in Lisbon shortly before setting sail:

If we can come to close quarters, Spanish valour and Spanish steel and the great masses of soldiers we shall have on board will make our victory certain. But unless God helps us by a miracle, the English, who have faster and handier ships than ours, and many more long-range guns, and who know their advantage just as well as we do, will never close with us at all, but lay off and batter us with their culverins, without our being able to do them any serious hurt.

This was why, as we shall see, the Spaniards did their best to make up for their disadvantage by careful preparations for a long artillery duel, and why the commanders could never feel satisfied even with the spectacular quantities of ordnance, powder, and shot they carried. Don Francisco de Bobadilla, Medina Sidonia's military adviser, later claimed to have predicted that the Spaniards would not be able to board and would therefore have to withstand an artillery duel for five days, 'and what would they do on the fifth day, with so little shot as they carried?'[3]

Just as the Spaniards knew that the English would stake all on firepower, so the English knew that the Spaniards would rely on grappling and boarding. English abhorrence of grappling during the Armada fight proved even greater than their customary cupidity for prizes. In the very first encounter, Juan Martínez de Recalde seems deliberately to have separated his ship from its companions to test the English resolve, without success: the enemy merely stood off and fired timidly at long range. Even when the *Rosario* of Don Pedro de Valdés was languishing, disabled, the English, at first, warily passed her by. On 2 August, the English let slip a genuine chance to board the *San Martín*, the Spanish flagship herself, when she was left an hour of sun away from the rest of the Armada. Though Medina Sidonia chivalrously signalled his willingness to try conclusions hand to hand, the English prudently settled for a long-range cannonade. According to the narrative of Don Juan Manrique, the *San Martín* returned the fire 'with so much gallantry that ... the enemy did not dare to come to close quarters with her'. The *Rosario*, the burned-out shell of the *San Salvador*, and the hospital ship *San Pedro Mayor*, grounded, after the campaign was over, on English shores, were the only prizes the English took; all were stricken ships, as was the galleass of Don Hugo de Moncada, at Calais—the only other vessel which the English, this time unsuccessfully, made any effort to seize. Neglect of prize-taking was so marked that it aroused suspicion at court. 'What causes are there', Richard Drake was charged to ask the Lord Admiral, 'why the Spanish navy hath not been boarded by the queen's ships? And though some of the ships of Spain may be thought too large to be boarded by the English, yet some of the queen's ships are thought very able to have boarded divers of the meaner ships of the Spanish navy.' It was, implicitly, to this kind of criticism that Sir Walter Raleigh replied in his later defence of English tactics. 'To clap ships together without consideration,' he wrote, 'belongs rather to a madman than to a man of war. For by such an ignorant

stayes and trimme her sailes, then it the secound
shipp to geve her fyre & the third and fourth &c
&c, wch done they shall all tack as the first
shipp & geve the other fyre keping the enimye
under a perpetuall volley. This yo must doe
uppon the leeward most shipp or shipps of an
enimye; wch yo shall either batter in peeces, or
force him or them to beare up, & so entangle
them & dryve them foule one of another to their
utter confusion.

The Musketeers divided in the
quarters of the shippe shall not deliver
their shott but at such distance as their
commaunders shall direct them.

If the Admirall geve chase & bee headmost
man, the next ship that follows him shall
take up his boate when shee is cast off, or if
any other ship bee appointed to geve chace
the next ship that follows her shall take
up her boate in lyke sorte.

If any man make any shipp to stryke sayle
hee shall not enter her till the Admirall be
come up unto him.

Yo shall take speciall care for the keping
of yo shipps & have betwene the decks & to have
yo ordnance in order & not cumbered wth trunks
or chests, but put their apparell in Cubbs Cloakbags

The evidence that line-ahead tactics, according to which ships would sail past the enemy, each firing a full broadside in turn, were known to the English at the time of the Armada, comes from these lines on sailing and gunnery discipline appended to William Gorges's treatise on defence of the Narrow Seas. The paragraph opens with

alsoe lay by y[ou]r fulls of water certaine y[ea]tt
blame... or th[orou]gh... to cast vppon and reabate any fyre
... ...

The master and Boatswayne in every ship shall
appoint a certaine number of saylo[rs] to eury sayle,
& to eury such companie a Masters mate, a boat-
swaines mate or quarter Master, soe at when eury
man knows his charge & place, thinge may bee
done w[i]thout noyse or confusion, & no man to direct
but these officers. As for example if the Master
or Masters mate bie heaue out the mayne topp
sayle, the masters mate boatswaynes mate or quart[er]
master w[i]th Eath[er] charge of that sayle shall
see his companie performe it so that onely one
to speake & w[i]thout tumult, And so accordingly for
all the rest of the Sayles. The boatswaynes
h[im]selfe takinge charge of noe perticuler Sayle
but overlookinge all & seeing eury man to doe his duty
in his place.

Noe man shall boord any enimyes shipp but by ord[er]
from a princip[a]ll Comander as the Admirall
vize admirall or Reare admirall, for that by one shipp[s]
boordinge all the fleet may bie engaged to honno[u]r
or losse. But eury Shipp that is vnder the lee of an
enimye shall labo[u]r to proue the ... if the Admirall
... it. But if wee fynde an enimye to leeward of vs
the whole fleet shall follow the Admirall vize admirall
or other leadinge Shipp w[i]thin Musket Shott of the
Enimye. Givinge so much libertye to the leadinge
Shipp as after her broadsyde is deliued, shee may

an injunction against precipitate boarding, followed by a recommendation that an enemy ship to leeward should be attacked with a series of broadsides 'keeping the enemy under a perpetual volley'.

bravery ... had the Lord Charles Howard, Admiral of England, been lost in the year 1588, if he had not been better advised than a great many malignant fools were, that found fault with his demeanour.' Even had the English wished to board, their ships were neither manned nor built for it. But they had no such wish: to remain beyond grappling distance was the essential means by which they hoped to frustrate Spanish tactics.[4]

The year before the Armada, a manual of Spanish Atlantic navigation and warfare was published in Mexico. The *Instrucción naútica para navegar* by Diego García del Palacio explains exactly what was meant to happen in such a fight as the Spaniards expected the Armada to have, in which a gunnery duel would be rounded off by grappling and boarding. It can be read today both as a means of evoking the atmosphere aboard a Spanish ship on the brink of battle and as a guide to the procedures the English knew their enemies would follow. When action was imminent, the gun decks were drenched with four fingers' depth of water 'because the chief thing that tends to cause damage and which must be taken into account aboard any ship of war is fire'. Shot was distributed to the guns, while small arms and grenades, stones and other missiles were piled on the castle decks; vats of vinegar and water, with cloths and sponges, were placed near the guns to cool them for reloading, and the masts and other exposed and inflammable equipment were girthed in soaking material. Diego García was emphatic about the need to marshal large numbers of musketeers and arquebusiers for the fight; the soldiery was to be used to help man the ship's equipment until their own arms of musket and arquebus came into use—particularly, again, with precautions against fire in mind. During the battle, the pilot would be in charge of the manning of the ship and the men in the rigging, on the watch, and at the tiller were under his orders. In a sheltered part of the ship a veteran gunner would be in charge of the powder store, watching against fire and passing out the parcelled charges, prepared in advance in the correct quantities for the small arms, with matches and fuses; it was also his responsibility to distribute extra powder for the big guns. 'The captain will also see', says Diego García, 'that behind cover in a secure place the surgeon shall be, with his brazier alight, and his instruments, oakum, eggs, turpentine, bandages, and two men of those least useful elsewhere, to look after the wounded who will be sent to him there.' The ships' boys had to help the gunners cool their pieces, and carry shot and wield damp blankets.

Half the ship's company was organized as a boarding party in two squads and armed with cutting weapons and pistols; the second squad was a reserve and could be committed if necessary; but on no account could the boarding party be further reinforced, so as not to endanger the defence of the ship. The pilot, taking the enemy ship—it was assumed—from a windward position, would heave to alongside, prow to poop. The enemy would try to luff, so as to fire a broadside at the approaching vessel; if he succeeded, the attacking ship would have to react quickly and do the same, but the normal approach would be bowsprit-on, and only the prow guns would come into play at the approach, first with ball, but then, as they closed, shooting at the rigging with chain-shot. A grappling hook would be fired by harpoon from the prow and another from the poop, secured by chains which could be easily loosed from the attacking ship: this was in case of fire, or failure, or of a hole opened between wind and water by enemy guns, or if the mainmast were felled by shot. Assuming an approach from the windward side, the attacking ship would be able to spatter the enemy with incendiary missiles. Meanwhile, the crew would take care to drench their own rigging and stand by to put out any fires the enemy might ignite aboard. The ships would be defended with a wall of netting strung over the bulwarks, which the boarding party would have to cut away while exposed to attack; partly to give time for this, the

Ceramic incendiary grenade from the wreck of the *Trinidad Valencera*; perhaps because shipboard fire-drill was so much practised on both sides, ship's fires seem to have neither disabled nor seriously damaged ships in conflict. Only the *San Salvador* succumbed to fire caused by an explosion in her powder magazine. But to set fire to enemy ships remained a constant tactical objective for which the weaponry carried by the fleets was designed.

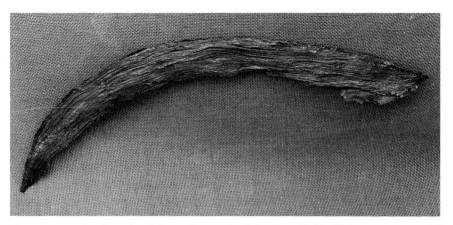

The remains of a shear-hook from the wreck of the *Gran Grifón*. This was an essential piece of apparatus for practitioners of boarding tactics, used to claw down the nets with which the enemy's decks would be protected against boarding.

attacking ship would have to rake the enemy deck with a hail of covering small-arms fire—'a continual storm of shot, not only because of the harm it will do through the gun-ports and arrow-ports but also because the smoke from our ship, which will be to windward, will not permit the enemy to see or to attend to any of their needs.'[5]

This was all very well, but Diego García did not explain how a determinedly elusive enemy could be brought to grapple against his will, especially when, as in the Armada fights, he almost always had the advantage of the wind. The worst fears of the Spanish commanders were soon fulfilled by the experience of the Armada. In the first exchanges, as Fray Bernardo de Góngora observed, the enemy 'having light ships with only sailors and gunners aboard, could open fire when he liked and withdraw when he liked'. After six days of almost continuous contact with the English in the Channel, Medina Sidonia complained that, 'our ships being very heavy in comparison with the lightness of those of the enemy, it was impossible to come to handstroke with them'. The duke offered almost the same words to Parma in justification of his request to borrow Parma's flyboats: but he was deluding himself; for all their lightness and manœuvrability, flyboats would have been too small to grapple the English galleons effectively. Every attempt to close was frustrated. When Martín de Bertendona gave them chase in the *Regazona* on 2 August, the English edged away. The galleasses put on speed with their oars, but still the English were too swift for them. On 4 August, the *Triumph* was becalmed and likely to be caught by the galleasses, but was towed away by her boats and 'wrested from our grasp'; later that day the duke's own ship, with a brief advantage from the wind, looked set to intercept her 'so that it appeared certain we would that day succeed in boarding them, wherein was the only way to victory. But at this moment the wind freshened in favour of the [*Triumph*].' Frobisher derided Drake's tactics in the great battle of 8 August; 'he came up bragging at the first indeed and gave them his prow and his broadside, and then kept his luff, and was glad that he was gone again'—but that was all any of the English ships were prepared to do as they sparred with the Spaniards up the Channel, stinging timidly and escaping with stings intact.[6]

Thus the English ships could not be pursued to close quarters, or induced to grapple by defiance. Yet, while the Spaniards were denied recourse to their own preferred tactics, they proved that they could frustrate those of the enemy by adopting a defensive formation which was

The *Ark Royal*, 'the odd ship in the world for all conditions', according to Lord Howard, had been commissioned as a private initiative by Sir Walter Raleigh as an affirmation of faith in modern galleon design, then acquired by the Crown in settlement of Raleigh's debts.

An old-fashioned English fighting ship, the *Jesus of Lübeck*, commissioned in the reign of Henry VIII, chivalrously decked out with banners and streamers—naval equivalents of the caparisons of a knightly steed. By the standards of Elizabethan ship design, her hull was stocky and her castles high: note the five banks of gun ports in the stern-castle.

effectively impenetrable by English methods. Essentially, this took the form of a convoy formation, first in a shape described by English onlookers as that of a 'crescent moon' with the horns, extending to the rear, formed of the strongest ships, and the supply vessels crowded into the centre, where the English could not pursue them without risking a general mêlée. On 1 August, Medina Sidonia modified this formation by ordering 'a continuous grouping to be formed of the van and the rear ... so as there should be no hindrance to our joining with the Duke of Parma'. In other words, the horns of the crescent were to meet, sealing and enclosing the hulks in a carapace of fighting ships; thereafter, the forty-three best ships in the fleet were always to be between the enemy and the rest, forming a wall of defiance rather like the face of a phalanx, beyond which the bulky non-combatants could crawl, untroubled, up the Channel. The admiral ordered that any captain who broke ranks should be hanged for it. The system was almost totally effective. On the English it made an initial impression of awe and an abiding effect of deterrence. 'We durst not adventure to put in among them,' admitted Howard, 'their fleet being so strong.' As he recalled towards the end of the campaign, 'All the world never saw such a force as theirs was'. Henry Whyte wrote to Walsingham of how, at first, the English were intimidated by 'the majesty of the enemy's fleet, the good order they held, and the private consideration of our own wants'. Sir Horatio Pallavicino, who arrived late and missed the first awesome sight, subsequently recalled that 'it was not convenient to attack them thus together in close order'. In the most determined probe the English made against the Spanish carapace, on 4 August, the defensive formation proved remarkably flexible: when the wingmost ship, the *San Mateo*, needed a respite from the English harrying, she was able to withdraw into the centre, to be replaced by the Florentine galleon, which saw off the attack.[7]

The English had no means to counter this tough and systematic tactical device. Their fighting style was chivalresque, or even Homeric, in which

While showing the familiar crescent formation, the Florentine ambassador's sketch of the Armada's proposed battle order has some curious features: the galleasses, which, in the event, occupied the wings, are shown as part of a central vanguard, preceded by four ships under Don Alonso de Leiva. The flanks are guarded by the *San Martín* and Recalde's flagship. The latter is clearly named as the *San Juan*, to which Recalde transferred at a late stage when his intended flagship, the *Santa Ana* proved unseaworthy: this may have been extraordinary ambassadorial prescience or a lucky mistake. The Florentine galleon is shown to the right of the main battle, marked 14.

ships would spar individually or fall in with others for concerted attacks as the occasion arose; but they started off the campaign without even an organization into squadrons to render tactical co-ordination possible. Their response to the Spanish formation was more or less instinctive: to keep the weather-gauge, to watch for and pounce on stragglers, and, above all, to maintain a dogged pursuit of the enemy, allowing him no respite in which to find an anchorage. The last aim was, to the English, the most important: it was Howard's declared aim at the outset of the campaign, and at its end the main ground of English self-congratulation: 'We have so daily pursued them at the heels, that they never had leisure to stop in any place along our English coast.' Yet, as we saw in the previous chapter, the Spaniards intended to move on anyway; the one moment at which the Armada might have tried for an anchorage, consistently with the decision of the council of war off the Lizard on 30 July, occurred at the eastern approach to the Solent, when it might have suited Medina Sidonia to try to seize the Isle of Wight. Even this, however, is very doubtful and it seems to have been the duke's intention all along to press on for an early rendezvous with Parma.

Opportunities for mopping up stragglers were few. Pallavicino claimed that 'if any ship was beaten out of the fleet she was suddenly surrounded and separated from the rest'; Howard boasted in a famous phrase that 'we pluck their feathers by little and little'; and the members of the Council consoled themselves that the Armada was 'daily weakened'. In reality, however, English cutting-out tactics were a failure. On the first day's fight, Recalde offered his own ship as a hostage to fortune; the English surrounded it but could neither capture nor seriously damage it; Recalde tried the same ruse—or perhaps this time the occasion was a chance one—with the best of the fighting hulks, the *Gran Grifón*, in the encounter of 3 August. Again the ship was extricated without mishap from the English trap. Two early prizes did fall to the English, but only by misadventures in which enemy action played no part: the *Rosario* of Don Pedro Valdés was disabled in a collision and placed beyond reach of a tow-line by a heavy sea; the *San Salvador* was set on fire by an explosion on board, when the fighting had subsided, caused by negligence or sabotage.[8]

In theory, the English could have used line-ahead tactics to destroy the most exposed ships. Orders appended to the manual known as William Gorges's 'Observations and Overtures for a Sea-fight upon Our Own Coasts', which were copied by Raleigh and which may have pre-dated the Armada, explain this system more or less as it came to be established in the days of Blake. 'If we find an enemy to be leewards of us the whole fleet shall follow the ... leading ship within musket-shot of the enemy.'

Each vessel was to discharge its broadside in turn, then 'give the other side, keeping them under a perpetual volley. This you must do to the hindermost ship or ships of the enemy, which you shall either batter in pieces, or force him or them to bear up and so entangle them.' Gunners were further enjoined 'not [to] shoot any great ordnance at other distance from point-blank range'. It is apparent from Sir William Wynter's account of the battle of 8 August that he may have been trying to put this tactical theory into effect, and there are strong indications that it was tried by Howard earlier in the campaign. On 2 August, for instance, the flagship of the Armada became detached from the rest of the fleet. By Medina Sidonia's account, the English flagship 'followed by most of his fleet'— and other accounts confirm that the English contrived to construct a very long line—passed by, unleashing her broadside, 'each ship firing at our flagship as it passed'. But the *San Martín*, which was strong in medium-range guns, returned fire and forced them to stand off where they could do no harm. It seems fair, at least, to say that Gorges's orders reflected, if they did not anticipate, the experience of the Armada. Certainly, the treatise to which they are appended is heavy with references to the Armada fight. The tactics they recommended, however, were patchily applied and disappointing in their effects.[9]

Only in the competition for the weather-gauge were the English uniformly successful. The surprise with which the English stole the wind on the night of 30 July comes through Pedro Coco Calderón's account; at nightfall, he wrote,

We descried the enemy fleet to leeward of us with sails furled but, because there was some mist and it was late, we could not make them out clearly. The duke ordered Captain Vicenzio that night to go through the Armada, giving the order to take up battle stations, for morning would break with the enemy upon us. The duke struck sail and spent that night lying in wait at anchor. When the moon broke through, which would be at about two in the morning the enemy hoisted sail and took the wind from us.

Thereafter, except for brief intervals when the wind veered round to the east, the Spaniards never had a chance of regaining the advantage of a windward position. It was universally assumed that a ship that had the weather-gauge could dictate the terms of battle. In the *Instrucción náutica* of 1587, for instance, it is assumed that, with the advantage of the wind, a Spanish ship would be able to catch and board an enemy. In the case of the Armada, however, this seems unlikely to have been possible. Even when a change in conditions briefly restored the weather-gauge to them, the Spaniards, in slower and heavier craft, had no hope of catching their elusive enemy. The 2 August, for instance, dawned with the Armada to windward: Don Martín de Bertendona seized the chance to turn on the pursuers, sailing hard for what he took to be their flagship, 'but she bare room and stood out to sea'. His Venetian argosy was in any case unlikely to match a purpose-built fighting ship for speed, though Bertendona may have deliberately tried to keep her light by sacrificing some guns, for she carried no truly heavy ordnance: if so, the sacrifice was in vain, for the *Regazona* was incapable of catching an English ship. On the 4 August, the two fastest vessels in the Armada, including the Castilian race-built *San Juan* of Fernando de Hora (see below, p. 243), tried to measure paces against English ships, but were left way behind. Even with their additional oar-power, the galleasses were unable to close with the English. Opinion of how they performed was divided after the event. John Montgomery, writing in 1589, was

credibly informed . . . that the said four galleasses made the most answer and offer of fight against us that was made, whereby they showed themselves to be ships of warlike force. For they having the vantage by their oars, might leave and take at their pleasure, and by reason thereof did often times issue forth of their squadrons as I may term it, as well to rescue their fellows sometimes distressed, as also to give charge upon some of ours, and then at their pleasure retired in again.

But this seems to mean that the galleasses were nimble by Spanish, not by English standards. Fray Bernardo de Góngora blamed their weight and wished for galleys instead, though he acknowledged the current caused them difficulties. They made another attempt in the calm of 4 August, but lost time towing away two threatened galleons, returning to the attack too late before the freshening of the wind. The English Admiral's summary may have done the galleasses less than justice: they 'singled themselves out from the fleet . . . but their courage failed them, for they attempted nothing'; Howard did acknowledge that they 'took courage' in the fight on the 4th. The effect, however, was the same, whether limited manœuvrability or lack of courage was to blame.[10]

The failure of the galleasses, the most agile Spanish fighting ships, illustrates the general predicament of the Spanish fleet. Though the wind favoured the English by keeping the Armada almost continuously on their lee, it was the inferior sailing qualities of the Spanish vessels that prevented the Armada from recovering the initiative. This was almost entirely a matter of ship design and of the composition of the fleet, rather than of seamanship. The general characterizations of all observers stress the handier qualities of the queen's ships. Elizabeth's galleons were 'the jewels

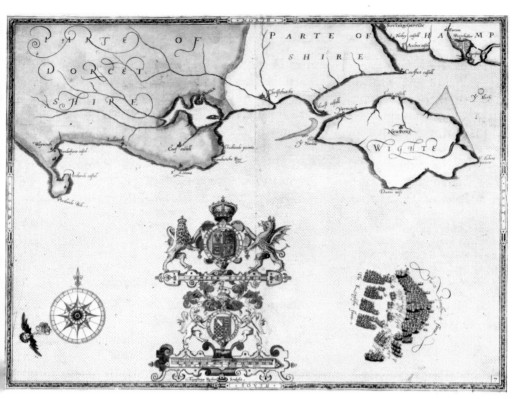

Robert Adams's version of the battle of 4 August shows a brisk exchange of ordnance. All the engaged English ships are being towed by their own boats. The *San Martin* is alone on the Spanish wing, while the galleasses nearby are operating by oar-power under bare poles.

of her kingdom', the 'noble ships' whose company Howard craved. The handiest of them, the flagship *Ark Royal* was 'the odd ship in the world for all conditions, and truly I think,' wrote Howard, 'that there can [be] no great ship make me change and go out of her'. Medina Sidonia praised the enemy ships as 'very agile and of such good steerage as they did with them what they wished'. Two of the Dutch deserters echoed a common opinion when they claimed that the Spanish ships handled badly and that it was easy for the English to take and keep the weather-gauge. Don Francisco de Bobabilla 'found the enemy with a great advantage in ships, better than ours for fighting, both in their build and in their artillery, gunners, and sailors, and rigged in such a way that they answered the helm and did all that was asked of them'.[11]

It is possible to get beyond these generalizations and to specify with some precision what made the English fleet easier to handle.

In the first place, the Armada included ships that were frankly unsuitable for northern warfare, or for any but routine use in Atlantic waters.

The Armada was delayed in Corunna because many of her hulks and 'Levanters'—the ships of Mediterranean build—had to run north before the storm until it was spent; it took a long time to gather them in, and they then caused further delays because some of them were apparently incapable of getting out of the harbour and round Cape Priorio without a free wind. Some of the Armada commanders were reluctant to be burdened with such unweatherly vessels. A letter of the king's to Medina Sidonia of 5 July 1588 reveals exactly the terms of the debate about the role of the Levanters and hulks within the Spanish camp:

I see plainly the truth of what you say, that the Levant ships are less free and staunch in heavy seas than the vessels built here [*that is, in Castile*], and that the hulks cannot sail to windward; but it is still the fact that the Levant ships constantly sail to England, and the hulks go hardly anywhere but up the Channel, and it is quite an exception for them to leave it to go to other seas. When they do it is for some reason other than the bad weather, or the working of the ships. It is true that if we could have things exactly as we wished, we would rather have other vessels but under the present circumstances the expedition must not be abandoned.

The king was right about the merchantmen being accustomed to the Channel; but to run with the wind on a peaceful voyage, with time on one's hands and no foe at one's back, was quite a different experience from that the Armada faced: running a gauntlet of gunnery while battling for the weather-gauge. The king was equally correct, and more to the point, in acknowledging the inferior build of most Mediterranean vessels. The shipwrights who made them were unused to Atlantic seas. They gave them slight frames in relation to their upper decks and the weight of arms they carried; their hulls were composed of more and smaller timbers than Atlantic-built ships. The ships, in consequence, were less stable and more fragile. The Armada voyage almost destroyed the Levant squadron: only twenty per cent of the Mediterranean-built vessels made it back to Spain, whereas of the galleons of Castile, which, of the Spanish ships, most closely resembled the English build, only one was lost. The king positively encouraged the Armada to sail with the deadwood of useless shipping. Even substandard ships would be worth taking, he told the duke, 'because the greater their number the more they will give pause to the enemy and prestige to the fleet, whatever eventuality may arise'. This argument ignored the tactical implications of the need to protect unserviceable ships from enemy action. The Armada was forced, by its own nature, into a defensive role. Even had the wind favoured the Armada in battle, even had the Spaniards possessed ships that could outsail the English, they could still move only at the speed of their slowest ship, and in battle could not risk exposing the vulnerable hulks that wallowed in the midst of the fleet.[12]

Even the best Spanish ships were laggard and clumsy in combat compared with their English counterparts. Contemporaries attributed this, almost unanimously, to a difference in weight. 'It is impossible to continue cruising with this Armada,' was Medina Sidonia's conclusion after struggling up the Channel, 'as its great weight causes it to be always to leeward of the enemy, and it is impossible to do any damage to him.' In the light of the experience of the Armada, Philip II was counselled on all sides to build lighter ships. This was the burden of Giulio Savorgnano's advice to his fellow expert, Filippo Pigafetta; indeed, in Savorgnano's submission, the ships should have been lightened even at the sacrifice of ordnance. Gian'Andrea Doria feared that unless lighter galleons were built, the fate of the Armada might befall the American bullion fleets. It is hard to evaluate this contemporary impression: calculations of tonnage were rough and ready, and the Spaniards and English used widely differing standards. It seems certain, however, that in terms of capacity and dead weight alike, the English fighting ships were actually, on average, bigger than those of their adversaries. On the other hand, the Spaniards were weighed down with a burdensome cargo of victuals and shot, with which the English could be supplied from on shore, and with all the extra soldiery and the impedimenta of land warfare for which the English had no use.

The more significant difference, however, was a matter not of dead weight but of build. In both fleets, of course, ships of widely differing designs were mixed, but in England, for longer and more consistently than in Spain, shipwrights had aimed to produce the relatively long, low, narrow, centrally masted, and therefore fast galleons that were best suited and best favoured for Atlantic conditions. While John Hawkins had been in charge of the queen's shipbuilding programme, this aim had been pursued almost fanatically.[13]

The beginnings of a low-built, weatherly, and gun-bearing fleet had been laid down in the reign of Henry VIII. But the *Mary Rose* was sunk, the *Great Harry* burned, and the *Grand Mistress* sold for scrap long before Elizabeth came to the throne. Towards the end of the reign of Mary and for the first decade of the new reign, a policy of restoring the navy was espoused, but it was only in the 1570s that naval shipbuilding began to be dominated by the galleon-type, long, low, and heavily gunned. The *Revenge*, of 450 tons, commissioned in 1575, was, whether as prototype or epitype, an influential ship, but also an expensive one, costing, by

Shipbuilding instructions from 'Fragments of Ancient English Shipwrightry' by Matthew Baker and others (1586). As in the contemporary Spanish designs of, for example, Diego García (p. 144), mathematical exactitude is given enormous emphasis. The complicated calculation in the margin explains a formula for working out the proper measurements without trigonometrical tables: 'On the other side', the heading reads, 'the way to find the lines is by help of the sines, but the same may be also found originally and is general for all cases in this manner.'

Hawkins's reckoning, nearly twice her proper price. For this and other irregularities in shipbuilding expenditure Hawkins blamed his own long-standing business partner, Sir William Wynter, whose 'covetousness' had been the cause of outrageous peculation—the queen's timber privately sold, for instance, and ships built at royal expense for Wynter and his clients. It does not seem that these accusations were taken seriously, but Hawkins was given the chance to make good his claim that he could have the work done more cheaply. On 1 January 1578 he was made Treasurer of the Navy, and in October 1579 he sealed his 'bargain' with the queen, by which ship repair was effectively to be farmed out to him and the professional shipwrights of the service, at his own cost, in exchange for payment of £2,200 a year. Not only did he save the queen money. He also took the opportunity to streamline the ships. William Borough's complaint that he had made them look like merchantmen by lopping off the fighting castles clearly evokes the spirit of Hawkins's reforms, The weight of professional opinion was on Hawkins's side, though Wynter counter-accused him of corruption and incompetence and Borough disputed the theory of the low-built fighting great ship. In 1585, Hawkins's 'bargain' was renewed and extended to cover the whole field of naval maintenance. Wynter and Borough themselves were silenced or converted when they saw Hawkins's vessels in action, and between 1587 and 1588 the three new galleons added to the queen's navy all reflected the values of the same school of naval architecture. The emphasis on the primacy of ordnance led necessarily to neater and therefore nimbler ships; big broadsides needed long flanks; doffed castles created dynamic profiles. As Raleigh pointed out, 'a ship of 600 tons will carry as good ordnance as a ship of 1200 . . . the lesser will turn her broadsides twice before the greater can wind once'.[14]

English ships, in short, were conceived as gun carriers, whereas those of Spain, while making big concessions to the importance of artillery, were still essentially troop-carriers, designed for the boarding tactics on which Spain relied. It is an instructive fact that the riskiest ship in the English fleet—the only vessel the Spaniards came anywhere near to catching, and that on two separate occasions—was the bulky galleon, *Triumph*, the most old-fashioned fighting ship in the queen's navy, which was encumbered with lofty troop-bearing 'castles'. All the Spanish fighting ships were like that. On the voyage out, the Portuguese galleon *Santiago* was given a new forecastle and higher, wider poop-deck to fit her better for boarding tactics. It was probably not so much the heaviness as the top-heaviness of Spanish ships that inhibited their handling. To make matters worse, in the search for speed, Spanish naval architects overcompensated for the undynamic profiles of their ships by trying to create more sailspace. Their vessels tended to be over-masted for their length, contributing to their top-heaviness; and although it was acknowledged, in Diego García's *Instrucción náutica,* for example, that the central mainmast worked best in the Atlantic, Spanish shipyards were still turning out vessels with the mainmast well forward, where the balance was disturbed. When Thomas Nash—introducing a theme often echoed ever since—described the Spanish ships as 'like a high wood, that overshadowed the shrubs of our low ships' he was trying to convey an impression of a conflict like that of David and Goliath. The implication was that only divine help could level the disparity, but, as in the original bible story, the natural advantages of a lean build and low profile were overlooked. David was deft, Goliath ungainly; a giant makes a big target; and, like the English ships, David was better equipped in long-range armament.[15]

Spanish tactics of relying on grappling failed because they could not be put into practice. English tactics of relying on fire-power were practised daily and intensively, but failed to work. Why was this? There can be no doubt that the English were liberal with their shot. After the first encounter, Howard wrote urgently to Walsingham for new supplies of 'some great shot of all bigness; for this service will continue long'. The first big battle on Tuesday, 2 August, off Portland Bill, was memorable, in Howard's own account, for the heat of the exchange of fire:

The fight was very nobly continued from morning until evening, the Lord Admiral being always [in] the hottest of the encounter, and it may well be said that for the time there was never seen a more terrible value of great shot, nor more hot fight than this was; for although the musketeers and harquebusiers were then infinite, yet could they not be discerned nor heard for that the great ordnance came so thick that a man would have judged it to have been a hot skirmish of small shot.

The battle of 4 August was described in remarkably similar terms by Sir George Carey, watching from the comparative safety of the Isle of Wight:

This morning began a great fight ... which continued from five of the clock until ten, with so great expense of powder and bullet, that ... the shot continued so thick together that it might rather have been judged a skirmish with small shot on land than a fight with great shot on sea.

Spanish veterans claimed that in some of the Armada fights they had endured 'twenty times as much great shot there plied' as at Lepanto; and the reports which reached Giulio Savorgnano, the Venetian engineer, suggested vividly, though with some exaggeration, that the English had kept up a 'continuous' cannonade for eight or nine days.[16]

These eloquent impressions notwithstanding, the damage done by English ordnance, prior to 8 August, was negligible. The only harm specifically recorded for this period on the Spanish side occurred, in the first day's action, to Recalde's rigging off the Eddystone on 31 July. The tally of a cut forestay and two great shot lodged in the foremast was after an hour's battering, which Recalde withstood alone, by three of the most heavily gunned ships in the queen's fleet, *Revenge*, *Victory*, and *Triumph*. The Portuguese turncoat, Emanuele Francisco, claimed rather more: that she was 'many times shot through', the prow and deck spoiled, the mast so weakened 'as they dare not abide any storm, nor to bear such sail as otherwise he might do'. But this is not borne out by the subsequent sailing performance of Recalde's ship, and may have been concocted by Francisco to please his English masters. Don Pedro de Valdés agreed with the official version: Recalde had been 'sore beaten' and 'his foremast was hurt with a great shot', but on the whole, the English gunnery had been ineffective.'There was little harm done, because the fight was far off.' Howard's assertion that he had made some Spanish ships make room to stop their leaks may be put down to wishful thinking. Over the next few days, the English did little better. Sixty Spaniards were reported killed in action on 3 August, out of a total of 167 lost prior to the Armada's arrival in Calais; some Flemish deserters claimed that the gunfire miraculously picked out the English and Netherlandish 'traitors' in the Spanish fleet. But there was no serious damage to ships. On 4 August, when the Spanish flagship, the *San Martín*, was pounded for an hour, the broadsides seem to have bounced off. The flagship lost only its flagstaff and one of the stays of the mainmast. The Spanish impression that the English gunners were 'so little skilled that of a thousand shot they fired scarce any hit our ships' was exaggerated, no doubt, but understandable. The English reflected a due sense of failure. When the campaign was over, the master gunner William Thomas, calling on Burghley to institute improved gunnery training, expressed the exasperation of a disappointed professional: 'What can be said but our sins was the cause that so much powder and shot spent, and so long time in fight, and, in comparison thereof, so little harm?'[17]

Spanish gunnery was even less effective than that of the English; nor

In this scene from the Armada playing-cards, the ace of clubs, with characteristic exaggeration, shows a ship sunk in the first action, whereas in fact on that day no damage was done and in all the fights only one ship sank as a result of the action. The English ships are shown by the cross of St George, while Spanish ships fly St Andrew's cross.

The Admirall ẏ L.ᵈ Sheffeild
S.ᵗ: Tho: Howard and others
joyn with Drake and Fenez
ag.ᵗ ẏ SpanifhFleet er worſt them.

did it improve, as that of the English did, towards the end of the campaign. The Spanish relations repeatedly record optimistic levels of damage to English ships, but they are uncorroborated by any reliable evidence. English admissions about casualties are equally incredible. After describing the intensity of the firefight of 4 August, for instance, Sir George Carey added the claim, 'in which conflict, thanks be to God, there hath not been two of our men hurt'. Yet, by any reasonable standard, the numbers of men lost in action must have been modest. Otherwise, Fenner could hardly write to Walsingham, with any prospect of being believed, at the end of the campaign:

God hath mightily protected Her Majesty's forces with the least losses that ever hath been heard of, being within the compass of so great volleys of shot, both small and great. I verily believe there is not three score men lost of her Majesty's forces. God make us as all Her Majesty's good subjects to render hearty praise and thanks unto the Lord of Lords therefore.

There were, however, enough wounded to divide between them a relatively large collective bonus of £80.[18]

Of damage to ships there is even less trace in the sources than of harm to men. Ubaldini's description of Drake's ship and cabin much pierced by shot, including saker and demi-culverin (ball of up to nine pounds weight), may be dismissed as one of the Italian's typical bits of romancing and as a rebuke to those, like Frobisher, who accused Drake of cowardice. Allowing for the exaggeration, it does appear that the *Revenge* had suffered more than most English ships, from an inventory made after the fight, which recorded Drake's vessel with a mainmast 'perished with shot as otherwise'. What is remarkable about the inventory generally, however, is the absence of serious damage anywhere. Apart from losses to rigging, only 'pestering' in the hull is widely recorded. There are no cases of major penetration by heavy shot. It is hard to believe that the inventory excluded hull damage mended by carpenters on board, for such repairs, though seaworthy, would have been seen as temporary.[19]

The lamentable performance of the Spanish ordnance can be understood, if at all, only in its proper perspective. Historians of the Armada have veered between equally unrealistic assessments of Spanish firepower, some regarding the artillery as a despised and neglected service aboard Spanish ships, others—since the research of Professor Michael Lewis revealed the previously unsuspected power and range of some Spanish guns—alleging that the Armada had more fire-power at its command than its enemies. In fact, the Spaniards were seriously outgunned, especially in certain crucial categories of ordnance; but their capacity was by no means contemptible. They knew they were in for a long and gruelling firefight, irrespective of whether a chance to board came off; and they were well prepared and well equipped to face it.[20]

The view one takes of the Armada guns depends on the sources one is contemplating. Actual guns—retrieved from wrecks and prizes—convey a different impression from that of the Spanish inventories. The former are solid evidence, but necessarily selective; the latter cover almost the whole fleet in detail but are dangerously unverifiable. Sixteenth-century inventories of war *matériel* generally incline to optimism, either because they were compiled by the suppliers or because they represented claims for reimbursement. On the other hand, in the last days of loading, in Lisbon and Corunna, the Armada can be supposed to have taken whatever guns were to hand, irrespective of what the inventories said; in some cases this meant that the tally in the inventory was actually exceeded in reality. The inventory of *Santa María de la Rosa*, for instance, lists only nine- and eighteen-pounders in the larger categories; but this is a wreck which has been excavated, yielding also five- and fifty-pound shot. Still, for a broad idea of the priorities and purposes with which Spanish ordnance was compiled, we are obliged to rely on the inventories. The English prizes—the *Nuestra Señora del Rosario* and the *San Salvador*—are too small a sample to be considered representative. The wrecks have yielded, on the whole, relatively small amounts of ordnance with a remarkable preponderance of very heavy pieces, which suggests that we are only glimpsing the unrepresentative residue of guns, most of which would have been jettisoned by ships in trouble. The inventories of what might be called the front-line fighting ships—the galleons and those armed merchantmen which are recorded as having contributed to the fighting—have been analysed by Dr I. A. A. Thompson, and the results compared with the breakdown of English ordnance analysed by Professor Lewis. In both cases, 'man-killer' guns of less than four pounds in weight of shot have been omitted from the analysis. Dr Thompson's sample of Spanish ships may not be completely representative: five merchantmen which actually had bigger guns than the Castilian galleons have been left out of the account; but while the precise figures may need some modification, there is no reason to question the validity of his method or the broad veracity and importance of his conclusion, which shows Spanish artillery to have been relatively lightweight and relatively deficient in ship-smashing guns. Less significantly, perhaps, from the point of view of effectiveness, the English were overwhelmingly superior in guns of greatest range.

The culverin was the gun of longest range, named for the length of its barrel, normally of at least ten or eleven feet and ranging, perhaps, up to twenty or more; the English reckoned pieces of this class to be eighteen-pounders, but all such figures have to be treated as approximations, as guns varied from maker to maker and piece to piece and one of the most crucial elements in the gunner's art was fitting the right shot to the right gun. Spanish culverins, for instance, were on average rather heavier than those of the English: by Dr Thompson's figures, whereas the Spaniards had only 37 per cent of English capacity in the culverin class by number of guns, they had 41.6 per cent by weight of shot. If one counts in demi-culverins, which had the same nominal range but fired a shot of only half the weight, English superiority at long range was even greater: 497 guns to 172. Officially, these guns were effective at a thousand feet: but that should be interpreted to mean that they could hit a large target on land at that distance—not necessarily that they would inflict any damage. These facts go a long way towards explaining the pattern of the conflict up to 8 August: the English were able to use their huge superiority in culverins and demi-culverins to maintain a cannonade at a safe distance, but without being able seriously to harm the Armada.

Taken further, the comparison shows also why the English were so tentative in closing the range, and that, when they at last did so, they chose to come as close as possible without risk of grappling. For the Spaniards did enjoy superiority in the medium-range heavy guns known as cannon and demi-cannon, with a nominal range of 800 feet. In this class, the English 'front line' had 275 guns against the Armada's 323. The culverins and demi-culverins could also, of course, play at this range, but English superiority in a medium-range fight was still only marginal in terms of numbers of guns. In practice, the English broadside at this range seems to have been much heavier-shotted, however, outweighing the Spanish by about seven to four. This is surprising, because the Spanish used the term 'cannon' to cover a wider range than the English, for whom it signified a gun normally of between twenty-nine and forty-four pounds, while in Spain it might cover a fifty-pounder. The inventories, however, divide the Spanish shot into fairly exact categories by weight and their testimony seems clear. Still, at medium range the fight would have been too nearly equal for English taste.

At close range, the Spaniards had a very large assortment of light-shotted weapons to bring into play. Their major ships had 347 eight-to fourteen-pounders, against 200 aboard the queen's galleons. These, however, were deck-clearing weapons, designed for the approach to boarding, as described by Diego García; none was a ship-killer, and many fired only stone shot which was virtually useless except against personnel. What counted at short range was the English superiority in heavyweight ordnance, a superiority of at least two to one by range and shot combined. For what they are worth, the figures for the full broadsides of both fighting forces are 14,677 pounds for the English and over 11,000 pounds for the Spanish. (Dr Thompson has pointed out that one can boost this figure by over a thousand pounds by reckoning from the capacity of the charges of powder used instead of from the weight of shot.) These figures are eloquent but crude: they obscure the fact that English superiority was concentrated in decisive areas.[21]

None of these calculations and speculations takes account of the fact that the Armada was carrying a huge and heavy siege train. Some of the fleet's working guns were forty-pounders—shot of that weight was inventoried aboard the *Santa María de la Rosa, San Salvador, Santa Bárbara,* and *San Bonaventura*. The siege guns, however, were much bigger. There were 48 fifty-pounders, some of which were aboard the *Trinidad Valencera*, discovered by divers from Londonderry in 1971; others went down off Tobermory with the Ragusan *San Juan de Sicilia* and have been known since the eighteenth century; these were among the finest guns carried by the Armada. The siege train also included heavy perriers, to judge from the six and a half inch stone shot discovered aboard the *Trinidad V*,*lencera.* There were a dozen even bigger guns, some of which Vicente Álvarez, a prisoner from the *Rosario*, thought were aboard his ship. These were probably not intended to be used at sea—indeed, they were probably carried unmounted. They were part of the siege impedimenta, which included brushwood for filling in moats, scaling ladders for assaulting walls, and mules for working the derricks. There is, however, one fragment of evidence which suggests that some of them may have seen action. When the Count of Fuentes was in Lisbon in February 1588, harrying the Master of Ordnance, he was told that forty demi-cannon and demi-culverins were even then being polished and mounted and that eight 'battery guns' and twenty field pieces had been embarked, which would be mounted when sea carriages were ready for them. Perhaps this promise was never fulfilled; if it was, it proved of no avail in the battles.[22]

The Armada's ordnance, though significantly inferior to that of the English, was obviously formidable enough. Why did it inflict neither serious damage nor heavy casualties? It has generally been supposed that there must have been some serious deficiency in the technical standard of Spanish equipment or gunnery technique. The guns were certainly of variable quality. Early espionage reports suggested to Walsingham that aboard the Armada 'the most part of the ordnance are cast iron pieces and lying very high, not to do any great harm'. The report was broadly right, as we shall see, about the positioning of Spanish guns, but the rest of the rumour rightly commanded little credit. The Spaniards despised iron guns, and, though their auxiliaries carried them, as did those of the English, they sought to avoid them for any but small, light pieces, like the obsolete wrought-iron breech-loaders, made of strips, hammer-welded, and bound with hoops, discovered by Dr Colin Martin aboard the wreck of the *Gran Grifón*. It was on bronze guns that the Armada relied and it would have been a serious matter if, as has been argued, a large number of them were rendered useless by defective casting. Dr Martin discovered a section of gun barrel so badly founded, honeycombed throughout, that it would have been in danger of exploding every time it was fired; in any case, as it was bored eccentrically it would have been incapable of accuracy. The wreck of the *Juliana*, excavated in 1985, has yielded another miscast gun, which exploded on firing. Yet these are obviously exceptional cases. Of thirty big Spanish guns excavated to date, or otherwise known, twenty-eight were sound. The quality of English casting was reputedly higher at the time than that of any other nation, and it is likely that in the English fleet the proportion of defective guns was even smaller. Evidently, however, the Spaniards had plenty of good weapons. Their own foundries were of little fame, but they exerted themselves to buy guns of German, Netherlandish, and English manufacture. It is by no means impossible that there was English ordnance aboard the Spanish ships. In 1574 a founder of Buxted had complained to Walsingham that 'there is often complaints coming before Your Honours about the shipping and selling of ordnance and cast iron to strangers to carry over the seas, they say in such numbers that your enemy is better furnished with them than the ships of our own country are'. The Spaniards, it was said in 1586, 'will give much for such a commodity'; the enemies of the realm were being allowed a share of 'a wonderful blessing of God, not proper to any other nation'. Raleigh claimed that 140 English culverins had gone to Naples; in 1587 nine shiploads of 'provisions, lead, powder, ordnance and muskets' had beaten the embargo out of Bristol, bound for Spain.[23]

The quality of the Spaniards' shot seems more questionable than that of their guns. Much of it was cast quickly, in response to the exceptional demand created by the Armada, and the founders may have sacrificed quality to economy. From aboard the *Santa María de la Rosa*, for instance, in 1969, Mr Sidney Wignall excavated shot so badly made that it would have crumbled on impact. A fifty-pound, seven-inch ball, for instance, had been cooled too quickly with cold water, becoming brittle in the process; the caster had saved more time by using an eccentric kernel of rejected three-inch shot, whereas the only good practice, as was well known at the time, was to cast a shot of one piece; when using small balls as core in their attempts to speed up the firing process, Spanish founders employed crushed haematite ore, mixed into the molten iron, to 'cushion' the kernels. The effect was to produce a shot considerably more fragile than the sturdy carapace of an oak ship. Again, however, as with the miscast guns raised by the marine archaeologists, it would be rash to suppose that these samples are representative of the Spaniards' store of shot in general. The Armada wrecks can yield only the worst shot, left unused after the battles, or consigned to the hold for ballast; and the very large shot intended for the siege train. Even so, the wreck of the *Gran Grifón* has been found to contain only good quality shot: the suggestion that this is because she was of German provenance and carried German shot is specious but unsound. As for powder, that was one commodity in which the Spaniards, for quality, were unsurpassed. The English could never obtain enough mined saltpetre for their needs: their best supplies came from Morocco, where, in the early 1570s, they began to trade in saltpetre for shot. Otherwise, they had to manufacture it by mixing earth with excrement, lime, and ashes, exposing it in a cold, dry place, watering it at intervals with urine and turning it many times. In consequence, their guns had to be primed with coarse powder, which needed a bigger charge—about a third as much again as fine powder—and produced a high rate of misfires. The Spaniards had no such problems. Consistently with the ideal recommended by Diego García, all their powder was of the fine-grained type.[24]

When Spanish guns, powder, and shot have been exonerated, gunnery practice is left to bear the blame for the failure of Spanish ordnance. The key to the investigation of this problem is evidence of the rate of fire. All witnesses agreed that the Spaniards could not remotely match the English rate; those who hazard a comparative calculation say that the English were three times as fast. The returns of equipment kept by the pursers of Spanish ships, however, suggest that even that figure may be flattering to the Spaniards. The *Trinidad de Escala*, for instance, managed no more than thirty-two shots, in all, on any one day; even in the terrible firefight of 8 August, when the written relations give an impression of innumerable blazing guns, she fired only thirty-one times. The *San Francisco* of the squadron of Don Pedro de Valdés seems to have been typical in having fired, on average, on the five days when there were major gunnery battles, one shot per gun per day. Indeed, it is hard to resist the impression that, for most of the time, most of the Spaniards did not reply to the English ordnance at all and that the greater part by far of the dense hail of ordnance we read of, for instance, in the accounts of Howard and of Sir George Carey, was fired from the English ships. This is a conclusion hard to reconcile, at first sight, with the Spaniards' frequent complaints that they were short of ammunition. As early as 4 August, Medina Sidonia wrote to Parma for powder and ball 'because with all this skirmishing I am running short'. But this was obviously an exaggeration as far as powder was concerned, and may be supposed to be so for ball as well, at least on the flagship which, by the official tally, had fired only a quarter of her store of shot by then. Fray Bernardo de Góngora reported the gunners' complaints that they were kept short of ball aboard the ill-fated *San Mateo* and *San Felipe*. Don Francisco de Bobadilla claimed it was lack of shot that silenced Spanish guns on 8 August and took the opportunity of congratulating himself on having predicted this outcome. Medina Sidonia, too, claimed that on that day there was no shot left aboard most of the front-line ships. We must also pay some heed to the Spaniards' common impression that they were firing quite freely, although not as fast as the English. Pedro Coco Calderón, for instance, thought the duke's flagship had fired eighty shots 'from one side only' in his hour-long duel with leading ships of the English fleet on 2 August, while Don Jorge Manrique put the figure at a hundred. The duke, more vaguely, concurred: his guns were 'well served and rapidly'. The flagship's official tally for that day was 120 rounds of great shot, all told. In the great battle of 8 August, Fray Bernardo exclaimed that 'there was no piece that did not fire at least a hundred times'. Unless this was unthinking hyperbole, he may have had the light ordnance and small arms in mind. The official figure for the great ordnance was 300 rounds, or just over six rounds per serviceable gun during probably about eight hours' close engagement with the enemy: in the conditions, that was by no means a contemptible rate of fire. On the other hand, the great guns seem to have been silent at some surprising times. In the dead calm of Thursday, 4 August, when Hawkins's ships were being rowed by their own boats against the Spanish vice-flagship, the Portuguese galleon *San Juan,* under Juan Martínez de Recalde, it took musketry to beat them off: this must be presumed to have had something to do with the English angle of approach, for surely, if the *San Juan's* guns could have been brought to bear, they would have fired. Captain Vanegas claimed that muskets had been brought to bear from Recalde's castles in the very first encounter—but he must have misremembered this episode: his account is quite at variance with all others, which uniformly stress the remoteness of the range.[25]

The contradictions of the evidence can be resolved if we assume that even the usual, sluggish rate of fire was sustained on only a few Spanish ships—the best and most active front-liners, those that ran out of shot in the midst of the critical battle of 8 August. Many Spanish gun crews, on many—perhaps most—of the ships, seem to have been almost incapable of reloading in action. If so, it may in some cases have been the result of incompetence or inexperience. In the Spanish service, unlike the English, the gun crews were not part of the ship's company. Small numbers of professionals, hired for the occasion or seconded from the land forces, were eked out with ordinary soldiers and some sailors, perhaps assisted by gromets. Many of the soldiers who joined the crews may never have fired a gun at sea. Those that had would in most cases have done so in the Mediterranean, where the role of the big guns would normally be to fire a single round at close range prior to boarding. The Spanish guns were kept loaded and primed between actions—Medina Sidonia's instructions are most insistent on that point; so the first shot was always easy to fire, if not to aim; thereafter, the job was hard—and evidently harder, in practice, on the Spanish than on the English side. It would, however, be unwise to suggest that Spanish gunnery practice was based on the expectation of firing only a single round or that on the Armada the gunners did not try to fire multiple rounds. The *Instrucción naútica para navegar* leaves no room for doubt that reloading in action was normal practice aboard the Atlantic warships of Spain; that it was done on the Armada is apparent from Pedro Coco Calderón's remark that he saw gunners killed and burned through not sponging their pieces properly between shots. It has been suggested—first, as far as I know, by the Armada wreck-hunter, Mr Sidney Wignall—that the Spanish gunners had to reload outboard. If so, it would be understandable, in the heat of a close firefight, if a gunner, obliged to dangle in the enemy's sights over a red-hot gun-barrel, were demoralized or killed. Professor Geoffrey Parker and Dr Colin Martin, who must be recognized as the leading authorities, have lent support to this theory by suggesting that the Spanish gun-carriages may not have been designed for inboard reloading. Those of the English clearly were: the English had four-wheeled truck-carriages which were suitable for the purpose at least from the time of the *Mary Rose*. The Spaniards certainly acknowledged the superiority of the English article and sought to copy it after the Armada. Yet, while the Spanish carriages were no doubt inferior, and helped to inhibit the rate of fire, they were probably capable of providing for inboard loading: the *Instrucción naútica*, again, assumes this and tells us that the gun decks were railed to help the gunners manhandle their pieces in and out. The suggestion that many Spanish guns may have been mounted on land carriages seems absurd; the Count of Fuentes's report shows pieces borrowed from on shore being converted to sea carriages, and while Juan de Vitoria may have been talking rather airily, as usual, in speaking of the artillery going aboard 'every piece with its sea carriage', he was probably substantially right. The Armada carried field carriages not for use aboard ship, but because they would be needed for the long land campaign which, it was hoped, would have to be fought in England. Such carriages have been found in the debris of the wreck of the *Trinidad Valencera*, where they were intended to accompany the siege guns stowed aboard. In some cases, in the desperate effort to increase the ordnance of the ships prior to sailing, field guns and siege guns were doubtless brought up on deck, and their carriages cannibalized. But only a small proportion of the Spanish guns are likely to have been mounted in this fashion. The siege guns were evidently not intended to be considered as part of the ships' complement of ordnance.[26]

The main outstanding problems—the low overall expenditure of shot, the low Spanish rate of fire, the varying rate of consumption of shot from ship to ship, and the failure of such shots as were fired to strike home or to effect much damage—become intelligible if considered in context without insisting on a single overall explanation. It must be remembered, first, that every time a Spanish ship engaged, it was in the hope of boarding. While English ships tried always to give their luff and fire, the Spaniards, to judge from the tactical orthodoxy prescribed in the *Instrucción naútica*, were approaching bowsprit-on, with only their prow guns able to play. Naturally, therefore, Spanish ordnance was little used in comparison with the English. Moreover, throughout the fighting, the brunt of the action was borne by only a few Spanish ships. In the first battle, on 31 July, Recalde's was the only ship seriously engaged; though others fired optimistically into the *néant*—Drake spoke of a general exchange of fire and the *Trinidad de Escala* recorded the expenditure of 22 balls, compared with 120 discharged by Recalde—most of the battle was out of the Spaniards' range. The next fight, on 2 August, impressed onlookers with the waste of shot, but it was again a single Spanish ship—this time the duke's flagship—which bore the brunt of the battle, firing 120 times. Most ships had no call to fire, and, when they did so, no chance to score a telling hit. On 3 August, it was the *Gran Grifón* that took the main part in a protracted artillery duel; though the flagship managed to expend 130 rounds, the account by Medina Sidonia implies that the damage done by the galleasses to the rigging of the *Revenge* was caused by a brief flurry of their great tail-guns. On the 4th, the action was more widely shared by Spanish ships; but, even so, only the galleasses with six of the great galleons and the *Rata Encoronada, San Juan de Sicilia, Regazona, Gran Grín,* and *Gran Grifón* made more than a token contribution to the firefight. When the relics of the Armada got home, there were many complaints 'of the

Detail of the barrel of one of the siege guns raised from the wreck of the *Trinidad Valencera*, cast by Rémy de Halut in 1556. The kingdom alluded to in the legend, 'Philippus Rex', is therefore England, of which he was king by virtue of his marriage to Mary Tudor. The arms of England are in the upper and lower quarters dexter of the coat of arms.

slackness of some of the ships'. According to Don Francisco de Bobadilla, 'the force of our fleet was a matter of twenty vessels; these have fought well and more than could have been expected. Most of the others fled every time they saw the enemy attack. This is not set down in the report out of regard to the good name of our country, but if you refer to Don Baltasar de Vanegas, he, as an eye-witness, will tell you what has happened.' The moral judgement has to be prised out of Don Francisco's statement and discarded: he was a landsman and may not have been fully aware of the genuine limitations of most of the ships. On the English side, too, it seems only to have been some twenty-seven of the best vessels that contributed directly to the fighting. It remains true, however, that on the Spanish side the ships of the most pugnacious or most conscientious captains were unduly prominent.[27]

This helps to explain why some ships had shot to spare, while others, in the front line, ran short at a critical juncture; Medina Sidonia busied himself with redistributing the shot during the voyage, but there were natural limitations to the effectiveness of such efforts. The English followed the same practice, concentrating shot in the great ships 'that thereby more service would be done.' Indeed, the shuttling of shot from ship to ship was standard practice, as essential as it was difficult. Fighting ships had to be kept light and a superfluity of shot could be an encumbrance; yet the fighting during the Armada campaign was, for those ships most hotly engaged, unprecedentedly voracious. If Fray Bernardo de Góngora was right in his belief that the gunners aboard the *San Mateo* and *San Felipe* had only the standard issue of 30 rounds per gun, that suggests that a good deal of the ships' share of shot was lying idle elsewhere. Both ships were inventoried for about 50 rounds per gun: The *San Felipe* with 2,000 rounds for 40 guns, the *San Mateo* with 1,700 for 34. Perhaps some of this was unusable because it was of poor quality or did not fit the guns. Yet it seems incredible that so large a stock could have been so quickly exhausted at the slow rates of fire that characterized the artillery of the Armada. It seems that on the Spanish side the means, as well as the brunt, of the battle must have been unevenly distributed.

The failure of English gunnery is, in a sense, even harder to explain than that of the Spanish. The English had invested enormous expectations in their fire-power, and uttered terrible threats. As early as 1570, a naval essayist was advocating artillery as an infallible winner of battles: 'For meeting with our Navy [the enemy] shall find nothing but blows, rending, tearing, sinking, firing or spoiling, for as much as he must come in the face of the cannon, in the eye of the ordnance.' The results of the Armada fights showed the hollowness of this sort of rhetoric. The firefight was the

English fleet's chosen tactical ploy; ship design had been subordinated to its demands so that all English hopes were staked on its outcome; the English had the benefit of relatively long and intensive experience of this type of warfare; their guns were the best available, their shot well cast, their sea carriages unquestionably handy, and their technical equipment in every respect superior, except for the coarse-grained powder they were obliged to use. It is possible, however, to identify two mutually linked reasons why, in the long period of sparring up to 8 August they scored few hits and none that truly told. In the first place, the range was too long, the fight 'far off', as Don Pedro de Valdés said. Occasional episodes were fought at closer range. On Tuesday, 2 August, Howard 'called unto certain of Her Majesty's ships then near at hand and charged them straitly to follow him and to set freshly upon the Spaniards, and to go within musket-shot of the enemy before they should discharge any one piece of ordnance, thereby to succour the *Triumph*', which at the time was in danger of being cut out by the galleasses; in fact, however, Howard's party was intercepted by the Spanish flagship and, despite the Lord Admiral's orders, seems to have lain off a long way to fire. The Spaniards did bring musketry into play on two other occasions, one on the Tuesday, the other on Thursday, 4 August, but these were probably brief incidents. Even for the heaviest ship-smashing ordnance, anything beyond musket-range must be considered a long shot. At that sort of distance, the 'face of the cannon' could be outfaced and the 'eye of the ordnance' outstared.[28]

In war, according to Napoleon, the moral is to the material as ten to one. Although little material damage had been wrought by either side during the days of sparring, it is worth while to inquire into the strengths and weaknesses of their respective morale. When the Armada was back in Spain and recriminations over its evil fate were rife, the king appointed Don Juan de Cardona to conduct an investigation into the state of the fleet and of its morale during the conflict, in particular. He unearthed accusations of cowardice and dereliction of duty, but claimed to have been able to discover nothing concrete about any particular case, save one, to which he attached enormous importance as the *fons et origo* of all subsequent difficulties. The incident to which he alluded was the most puzzling of all the days of sparring, the loss of the flagship of the Andalusian squadron, the *Nuestra Señora del Rosario* of Don Pedro de Valdés, which was immobilized by an accident on 31 July and captured by Drake on 1 August, without a fight. The *Rosario* was of Galician build, called a 'carrack' by the English, big and broad-beamed. She was one of the most powerful and valuable vessels in the Armada, with 46 great guns and 50,000 of the king's ducats aboard. The questions her loss posed were:

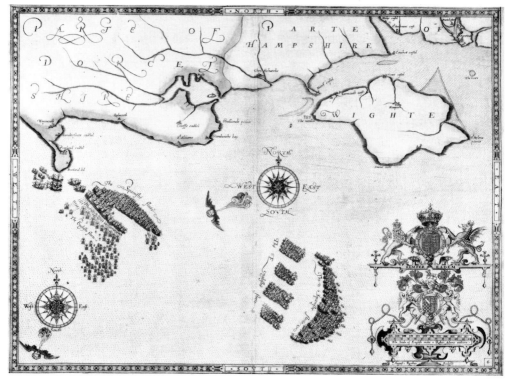

The battle of Tuesday 2 August. The battle for the *Triumph* is shown at the extreme left, with the galleasses trying to use their oar-power to catch her. The Spanish main battle seems to be bringing its broadsides to bear against the English. Whereas the English attack is made in skirmishing order, the new, well-ordered formation into squadrons decreed by Howard next day is shown in the centre foreground.

why was she abandoned by the rest of the fleet? Why was she yielded so easily to the enemy? And what was the effect of this apparently craven and bungled episode on the morale of the Armada? Don Juan de Cardona's findings, though they need to be corrected by comparison with other sources, make a useful starting point. 'The principal matter', he wrote,

was the loss of the ship of Don Pedro de Valdés, and about that I have tried to get special information, as I think it was the beginning and cause of the rest. The ship of Pedro de Valdés was sailing ahead of the Duke of Medina Sidonia's galleon, and to one side of it, going up-Channel, with the enemy fleet following them. Valdés's ship fouled another ship of its own squadron [*identifiable from other sources as the Andalusian* Santa Catalina], breaking its bowsprit and foremast, which fell on the ship and struck the mainsail, taking the wind out of it so that it could not proceed, but fell behind the other ships and galleons, which did not help it or show signs of wanting to. In this way it passed near Medina Sidonia's galleon, which likewise showed no sign of helping it, nor did it haul to, or take in any sail, or make signal for the fleet to wait. As Valdés's ship was falling behind, and Medina Sidonia's and the fleet advancing, the duke asked Diego Flores de Valdés if help should be given. Flores answered that Valdés was of his own blood and his friend, but that more was owing to the service of His Majesty. If the duke got into difficulty helping the ship, he would be lost and imperil the whole fleet; this is the same information as the duke gave Your Majesty. Don Jorge Manrique said that assistance should be given to Valdés's ship, and that not to do so was to lose both honour and the fleet. He offered himself to do it, but the duke replied that he would follow the advice of Diego Flores, who was the one Your Majesty had sent to him. Thus Valdés's ship was lost, without any ships going to him except a few small ones sent to see what they could do, which was nothing; it was lost in sight of the whole fleet, which must have been a great grief and injury, for the enemy was encouraged and Your Majesty's fleet disheartened, and the murmur went round the fleet, let no ship get into difficulties, for if they have not helped the chief of the squadron, whom will they help if he gets into danger?[29]

Medina Sidonia seems to have sensed some damage or potential damage to the fleet's morale, especially as the loss of the *Rosario* was compounded by the explosion aboard the Guipúzcoan *San Salvador*. He responded—as he would again next time he felt morale was ebbing, when the Armada was in the North Sea—by a terrible admonition designed to enforce discipline: pinnaces toured the fleet, with gallows erect and hangmen at hand, circulating orders that any departure from formation would be punished by death.

In one respect, Cardona's report is reticent. He is scrupulous to avoid the suggestion, which was made by Don Pedro de Valdés himself, and may have been current in the fleet, that the duke and his adviser were motivated by personal hostility towards the man they abandoned. Don Pedro claimed that Medina Sidonia had 'looked on him with an unfriendly eye and had used expressions towards him which grieved him' after the council of war at Corunna on 27 June, at which Valdés, alone of the commanders present, had urged immediate departure for English waters, without waiting for the Armada to regroup after the storm that had dispersed the fleet. He found his advice spurned, 'a good occasion lost', and his commander 'using me differently from the other generals'. Despite the assurances Cardona reported, Diego Flores was notoriously on bad terms with his kinsman—sufficiently so to fuel the rumour that he had deliberately contrived his downfall. In a vivid image, Juan de Vitoria likened Diego Flores's advice in favour of the abandonment of Don Pedro to Caiaphas's precept (John, 18: 14), *Unus homo moriatur ne tota gens pereat*.[30]

In other respects, however, Juan de Cardona seems to have been influenced by the febrile atmosphere of suspicion and reproach that prevailed in the fleet in the aftermath of failure. It was not strictly true that Don Pedro was lost 'in the sight' of the entire fleet. Some crews may have suspected, as they sailed away, that they had seen the last of him. A natural assumption, however, was that he would be towed out of danger and effect repairs before proceeding in safety. Cardona omitted, moreover, the natural explanation of the predicament which led to Don Pedro's plight, though it is mentioned in other sources: the sea grew particularly heavy at the time of the attempted rescue of the *Rosario*, and it proved genuinely impossible to get a line aboard her. Fray Bernardo de Góngora, whom Don Pedro sent to the duke's flagship to ask for help, declared explicitly that Valdés's misfortune was caused by a heavy sea, though he added, 'The duke wished to help him and Diego Flores insisted that he should not do so and should not put our Armada at risk'. Medina Sidonia's own account is consistent with Fray Bernardo's, and all others, which stress the adverse weather. The heavy wind and high sea frustrated his own efforts, from the flagship, to pass Don Pedro a line. Diego Flores warned that to shorten sail would imperil the Armada, as the rest of the fleet would lose sight of the flagship. Four pinnaces, the leading Castilian galleon, and a galleass were then detailed to succour the *Rosario* but the bad weather and the onset of darkness prevented them either taking the ship in tow or transferring her personnel. Broadly, this is confirmed by English witnesses aboard the *Margaret and John*, who claimed to have driven away three pinnaces left to help Don Pedro.[31]

Valdés compounded his misfortune the following day by his abject surrender to Drake. Juan de Vitoria assumed that he must have fought to the last handful of men and 'done his duty'; indeed, it would have been his duty to resist capture with every means at his command. In the scale of values of his service and his time, it was better to go down fighting than to grant the enemy a prize, especially one that brought with it a fortune in the king's gold and a formidable accession of strength in the form of great brass guns and corresponding quantities of fine powder and shot, which would certainly be used against his comrades of the Armada if they fell into enemy hands. To resist with the tenacity of, say, a Richard Grenville of the *Revenge* was supererogatory, but Don Pedro should have fought to expend his own powder and shot and, if possible, scuttled his ship. When Don Hugo de Moncada was grounded at Calais a few days later, or Don Felipe Pimentel at Nieuport, no thought of surrender entered their heads until every resource had been exhausted. Don Pedro's resources included 46 great guns, 114 quintals of powder, and 2,300

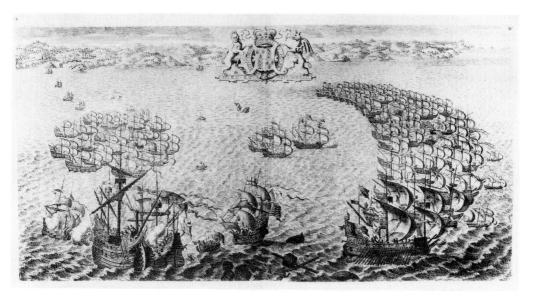

Though they survive only in early eighteenth-century engravings by John Pine, the House of Lords tapestries depicting the Armada conflicts are a source of enormous value, completed and hanging in Arundel House by not later than 1602. The scenes are closely related to those selected by Robert Adams for his charts of 1588, but the tapestry engravings add considerably to those prototypes in matters of detail. This scene, however, of the capture of *Nuestra Señora del Rosario* is characterized by considerable artistic licence, as the English did not grapple or board in action, and the capture was delayed until the day after the *Rosario* disabled herself, when the rest of the Armada was out of reach.

rounds of heavy shot. Only eight or nine ships of the Armada were more powerful than the *Rosario*. Her spoils were ideal to answer England's needs. When the *Salvador* came to port, the mayor of Weymouth acknowledged Howard's order 'that all the said powder and shot be sent unto his Lordship with all possible expedition, for that the state of the realm dependeth on the present supply of such wants'. To surrender these supplies to the enemy doubled Spain's loss. Don Pedro's excuse was that it was no dishonour to surrender to a foe as formidable as Drake, who was endowed with 'valour and felicity so great that Mars and Neptune seemed to attend him'. The most that can be said on his behalf is that he seems to have assigned a higher value to discretion than to valour.

The psychological and material effects of the loss of the *Rosario* were exacerbated by the mysterious explosion that incapacitated the Guipúzcoan *San Salvador* on the same day. The problem of whether accident or sabotage was to blame is insoluble; the damage, however, can be nicely calculated. The *San Salvador* had one of the biggest powder magazines in the fleet: indeed, at 130 quintals she was exceeded only by the three most

Nearly a hundred years after the event, the Armada playing-card achieves a more realistic version of the loss of the *Rosario* than the near-contemporary image-makers, but with the sacrifice of drama.

The Galleon of Don Pedro taken Prisoner by Sʳ. Francis Drake, and sent to Dartmouth.

heavily gunned ships, and carried far more powder than was needed for her twenty-five guns. Little or none of this powder can have been saved; on the other hand, little can have been retrieved by the English, and the loss to the Armada was bearable as powder was held in ample quantities throughout the fleet and no ship reported a shortage even after the heaviest fighting. As well as being a munitions vessel, the *San Salvador*, like the *Rosario*, was a pay ship; the paymaster, however, 'with a part of His Majesty's treasure', escaped the wreck. More serious was the loss of men in horrifying circumstances. The spectacle of the explosion, and the recovery from the sea and the wreck of men suffering from burns, must have had a daunting effect on the rest of the fleet; what was perhaps worse was that the heavy seas in the evening, which impeded the rescue of the *Rosario*, also interrupted the evacuation of the *San Salvador*; the next morning the work could not be resumed, with the English drawing on; and some of the most severe cases of burns had to be left on board to be captured by the enemy. Pedro Coco Calderón has left the most graphic eyewitness account:

The two poop decks were blown off with more than two hundred men on them, including the ensign-bearer Castañeda, who was guarding the magazine. The ship split afore and astern. Many men were flung into the sea where they drowned, but most of the company were saved in four pinnaces sent by the duke; among them was the paymaster Juan de Huerta with his staff and records and some of the money in his care. The ship sailed on, albeit laboriously, all that day and night, until on Monday morning, first of August, the duke ordered the remaining crew to be transhipped and the vessel to be scuttled, and as the captain was badly wounded, and the seamen were the first to be rescued, there was no one left aboard to scuttle her. Moreover, many burned and wounded men were left, who could not be rescued because the enemy was at hand. It is believed that the enemy will have taken her in tow and got her to some harbour on their coast. And in the vice-flagship of the hulks, I, Pedro Coco Calderón, purser to the Armada, took aboard Captain Villaviciosa and no fewer than ninety-four men suffering from burns.

Two or three companies of soldiers escaped unharmed, but the ship's master came off with burns and Captain Pedro de Pliego 'had burns all over'.[32]

The material loss to Spain was serious; the weakening of morale, though perhaps exaggerated by Cardona, may well have been significant. Yet, strange though it may seem, the episode had a comparably deleterious effect on the English fleet. The circumstances in which the *Rosario* was captured provoked, on the English side, almost as much dissatisfaction and dissension as were aroused on the Spanish side by the circumstances of her loss. In the Armada, Don Pedro and Don Diego were the antagonists around whom opinion clustered; in the English fleet, the corresponding roles were filled by Frobisher and Drake. Drake's alleged transgression was twofold—an act of indiscipline and theft; theft in arrogating the prize of Don Pedro's ship to himself, and indiscipline in breaking ranks to do it. Philip II had inveighed menacingly against commanders who were 'motivated by greed' and who 'give pursuit to the enemy and take prizes' when they should have been maintaining, in formation, the collective objectives of the fleet. English standards were more lax, perhaps, but Drake was aware of his responsibility, especially as, on the night the *Rosario* was disabled, he had been detailed to lead the pursuit of the Spaniards. His orders were to show a lantern in his stern that would guide the rest of the English navy. Yet in the dead of the night he extinguished the lantern and put about secretly, leaving other ships to manage as best they could. The excuse he offered subsequently was that, misled by the passage of some innocent merchantmen, he suspected that the Armada was stealing its way round to the rear of the queen's fleet in order to hold the weather-gauge at daylight. In principle, this was not impossible: the Marquess of Santa Cruz had effected a similar stratagem at the battle of Terceira, with decisive effect. Yet the tale as Drake told it defies belief. No other vessels in either fleet sighted the alleged merchantmen. And Drake's claim to have discovered the *Rosario* by accident the following day strained the credulity of his audience. From this moment on, the animosities that divided the English command were ill-concealed. A witness reporting what he claimed were Frobisher's complaints, seems to have reproduced their authentic timbre. Drake boasted, according to Frobisher, that 'no man hath done any good service but he' and

he hath done good service indeed, for he took Don Pedro. For after he had seen [*the* Rosario] in the evening, that she had spent her masts, then, like a coward, he kept by her all night, because he would have the spoil. He thinketh to cozen us of our shares of fifteen thousand ducats; but we will have our shares, or I will make him spend the best blood in his belly; for he hath had enough of those cozening cheats already.[33]

The impact of the affair of the *Rosario* broke on forces which already showed signs of selectively or patchily weak morale. As well as by dissension and indiscipline, low morale can be measured by desertion, and although desertion was almost impossible at sea, there is evidence of the fear of it, especially on the Spanish side, prior to sailing. On the voyage, of course, the discontent that breeds desertion must be supposed to have sought other and perhaps more harmful outlets. While laid up in Corunna, Medina Sidonia issued shore passes only for the purchase of food and water for fear of desertion. He kept 'the closest possible watch on our soldiers and sailors, guards being posted all along the shore, and in the roads and passages by which they might escape'. On the other hand, these measures seem to have been effective: no soldiers and only a handful of sailors were reported absent (though it must be remembered that 'dead men's pay' always kept muster-lists artificially high). Medina Sidonia, moreover, was speaking of a peculiarly difficult period for Spanish morale, when the Armada had been dispersed by a storm of apparent ill omen, and on the very day of his most pessimistic report, 6 July, he was able to record an improvement. Thereafter his fears were centred on the unreliability of the Armada's Portuguese contingent. He had mistrusted them from the first. They were 'not to be counted on'; their captains had refused their commissions; loyal Portuguese were disgusted by the poor response to attempted recruiting. Gregorio de Sotomayor, a prisoner taken by the English aboard the *Rosario* claimed that 2,000 Portuguese had been compelled aboard the Armada on pain of death and responded to interrogators with the claim, 'as a good Portingal ... I would be glad that I knew more, both for the service of Her Majesty and preservation of this kingdom'. Other Portuguese prisoners made similar efforts to distance

Arthur L.ᵈ Grey, S.ʳ Francis Knolles, S.ʳ Iohn Norris, S.ʳ Richard Bingham, S.ʳ Rog Williams & others in a Councell of war, consulting how ÿ land Seruice should be Ordered

The 'land men': the artist of the Armada playing-cards must have been influenced by a document much copied in the seventeenth century, perhaps because it was felt to recommend useful precautions against the invasion: 'The opinion of Ye Lord Gray, Sir Francis Knowles, Sir John Norris, Sir Richard Bingham, Sir Roger Williams and others, what places were most likely the enemy would land at and what were meet to be done to make head against him.'

themselves from their Spanish comrades, some of them literally, it was claimed, by swimming to English ships during the fighting. Their protestations were believed in England, where men were predisposed to doubt the usefulness of heterogeneous forces, because of the Renaissance doctrine that the most effective armies were motivated by the collective self-interest of fellow citizens. According to Ubaldini, Queen Elizabeth preferred to exclude foreign ships and men from her service for the sake of a 'common tongue' and 'to encourage those who (besides considering the common honour and that of the crown) would have a good care of their properties, of their women and children and lastly of their own native land'.[34]

It must be doubted, however, whether there was in practice much difference in the level of commitment of both forces. Another prisoner, Gonzalo González del Castillo, a native of Granada, who was captured on the hospital ship *San Pedro Mayor* and spent three years in England, took the common-sense view that there were subjects of both crowns—English and Spanish—who were affected to the other side, generally for reasons of religion or war-weariness. The English land forces, being less professional, were more prone to desertion and more susceptible to fear than their

Spanish counterparts. Howard thought 'a little sight of the enemy' would 'fear the land men much'. Morale is a phenomenon of mass psychology and there seems, in most conflicts, to be a variable threshold beyond which panic becomes irreparable. In many battles, the outcome has depended on the relative distance from that threshold of the respective contenders. In the Armada campaign, the English collectively, for all their deficiencies of discipline, never seem to have got anywhere near it. The Armada, on three occasions, was to draw perilously close: first, on the night of 7 August, when the formation of the fleet was broken by its headlong flight from the fireships in Calais roads; then again on 10 August, when some ships broke ranks and Medina Sidonia responded with Draconian measures to prevent a recurrence; and finally, on the way home, amid the terrible storms of late August and September 1588, when the Armada virtually broke up—even before its final dispersal by intolerable weather—and sought safety in a spirit of *sauve qui peut*. Each of these episodes must be examined in its place. But although each imperilled the morale of the fleet, in none did it break down irretrievably; nor can any of them be said to have made a decisive difference to the outcome of the campaign. Though the days of sparring frayed the nerves of both sides, the morale of neither was seriously corroded, even by the ominous circumstances of the surrender of Don Pedro de Valdés. While Juan de Cardona's report may contain a grain of truth, and the men of the Armada must have been perturbed by what they had witnessed, the advantage, as they approached the Narrow Seas, was finely balanced between mutually wary but undaunted foes.[35]

Seven

———— ▸✖◂ ————

The Face of Battle

DURING the days of sparring in the Channel the English made no impact on the Armada. There is some evidence, however, that they were consciously experimenting with range: on 2 August, as we have seen, there was Howard's order, frustrated, perhaps, in the event, to close within musket-shot in the attempt to rescue the *Triumph*. On 4 August, Medina Sidonia had the impression that the English were probing closer than before, though they drew back again in the event. Henry Whyte had the impression that as he and his comrades became emboldened by the passage of time and the arrival of their reinforcements, their attacks gradually intensified, though he does not specifically speak of closing range. It seems safe to infer, however—particularly in view of what was to happen subsequently—that by the end of the battle of 4 August two inescapable conclusions had impressed themselves upon the English: first, that their ordnance would not yield satisfactory results except at short range; secondly, that only the breakup of the Armada's strong tactical formation, or a formidable accession of new strength to their own side, would give them the opportunity to do so in safety. Within the next few days, both conditions came to be fulfilled.

There was no action on 5 or 6 August. This may mean simply that the English felt cowed; or it is equally possible, though none of the sources actually says so, that they were steeling themselves for a further, more purposeful trial of strength. The munitions they had were conserved and supplemented with whatever could be got from on shore. All Friday and Saturday, 5 and 6 August, according to the official English report, 'the Spaniards went always before the English army like sheep, during which time the justices of peace near the seacoast, the Earl of Sussex, Sir George Carey and the captains of the forts and castles alongst the coast, sent us men, powder, shot, victuals and ships to aid us'. This showed the advantage of fighting the Spaniards in English home waters. Had the seamen's strategy been followed of seeking the Armada on the Spanish coast, the parts of the antagonists would have been reversed and it might have been the English who were caught short of shot at a critical juncture. The principal accession of English strength came from the junction of the main fleet with Seymour's force, which had been guarding the Narrow Seas against a sortie by the Duke of Parma. Back on shore, fear of Parma was reaching the proportions of panic. On 5 August, for instance, Leicester was afraid that 'the Prince is looked to issue out presently'. According to Sir Roger Williams's information, 'he hath suffered no stranger this seven or eight days to come to him, or to see his army and ships, but he hath blindfolded them'. If the Armada reached the Narrow Seas, 'the Prince will play another manner of part than is looked for'. From Seymour's perspective, however, the watch on the Flemish coast seemed pointless: he and his men were being excluded from the honour and glory of the main fight in order to blockade an enemy who was most unlikely to attempt anything. In a letter to Walsingham on 6 August, Seymour's confederate Sir William Wynter advised the withdrawal of the Flanders guard and a general concentration of shipping at the Norehead: that may not have seemed a very sensible suggestion, but it illustrated the impatience of Seymour's men at their enforced passivity. The following day, the Council gave in to their entreaties. 'Her Majesty seeth', they wrote to Seymour, 'how greatly it importeth her service to have somewhat done to distress the Spanish navy before they shall join with the Duke of Parma's forces by sea.' Seymour's ships—thirty-five sail, including five great galleons—were ordered to leave their blockading station and join the Lord Admiral. It does not appear that in taking this decision the English were relying on their Dutch friends to keep up the watch against Parma. The Dutch had been reported as having put into harbour three days previously on account of bad weather; as far as the English were aware, there were no Dutch vessels abroad. Dutch sources bear out their later claim that they were back on the watch by the time of Seymour's departure or soon thereafter. But at the time no one in England or in the main English fleet knew where they were or what they would do or, even if their co-operation could be relied on absolutely, how effective they would be. In any case, Seymour seems to have anticipated his orders. He was with Howard for a council of war early in the morning of Sunday, 7 August. The English fleet now, by Howard's reckoning, amounted to 140 sail: Seymour's arrival ensured that in the next encounter, moreover, the English would have a clear advantage in first-rank fighting ships, for the first time since the pursuit of the Spaniards began.[1]

Once Medina Sidonia had decided in favour of the plan of meeting Parma as near as possible to Dunkirk, and escorting him across the Channel, the choice of Calais as an anchorage was more or less forced on the Armada. It was as near as the Armada could get to the Army of Flanders. No other deep-water port was available. 'There was some difference of opinion', says Medina Sidonia's report, 'as to whether we should anchor here, the majority being in favour of sailing on. The duke, however, was informed by his pilots that if he proceeded any further the currents would force him to run out of the Channel into Norwegian waters, and he consequently decided to anchor off Calais.' This 'difference of opinion' is hard to interpret. What alternative can the duke's subordinates have been advocating? We can be reasonably certain that they were not sustaining or reverting to the discarded plan of a rendezvous with Parma 'off the Cape of Margate', first because, as we have seen, all mention of that plan had been dropped since before the voyage of the Armada began; secondly because what I have called the plan for a convoy, involving as early as possible a rendezvous with Parma, had been implicit in all accounts of the course and deliberations of the Armada up the Channel; thirdly because it is incredible that anyone in the fleet could have believed that Parma would venture on a Channel crossing on his own: fatally bad as were the communications between Flanders, Madrid, and the fleet, that message, at least, had got across. Thirdly, when the weather at last forced the Armada into the North Sea, the duke's council of war continued to profess the hope of returning to the Channel to effect the junction with Parma. The possibility of doing so in the Thames estuary does not seem to have been mentioned. Finally, and, I think, conclusively, we know by his own account that Don Francisco de Bobadilla was one of the officers opposed to a halt at Calais: yet it is equally clear that throughout the voyage he had been expecting to meet Parma as early as possible, and as close as possible to Dunkirk. There can therefore only have been three motives for advising the duke to sail on: the hope of bringing the enemy to battle at last—surely forlorn by now but redeemed, perhaps, by the belief that the English might be emboldened by their growing strength; the chance—surely remote and risky—that a change of wind might favour the Armada; and the desire to abandon the enterprise altogether. That this last motive was rife was later confirmed by Don Jorge Manrique, who admitted 'that some advised the duke that, instead of anchoring, he should sail through the straits of Calais and return by the north of Spain, and that he opposed this and ordered them to anchor'.

Certainly, Calais was far from ideal. 'I do not know', wrote Don Francisco de Bobadilla, 'whose idea it was that in a port with such strong currents and on an open shore with cross winds and so many shoals from one side to the other these forces could join, for it is not a port in which one could delay without danger.' Medina Sidonia interpreted in a friendly sense the governor's warning 'that the place where he was lying was extremely dangerous to stay in, in consequence of the cross currents of the Channel being very strong', though it seems more likely to have been a polite attempt to get rid of an uninvited and unwelcome guest. But it was no more than the truth and Medina Sidonia was already aware of it, from the information of his expert pilots, before he came into the roads. As soon as he arrived he wrote to Parma 'to inform him of the position of the Armada, and to say that it was impossible for it to remain where it was without risk'. The French attitude, if hostile, could have made Calais an untenable berth. Drake somehow got the idea that the mayor of the town inclined to the English and that 'Her Majesty is beholden unto' him; in fact, however, the French seem to have favoured the Armada as much as was consistent with their neutrality. The governor of the town gave Medina Sidonia 'a great present of fresh provisions' and allowed him to buy more ashore, thereby enriching his fellow citizens by six thousand ducats. This must have been a welcome relief to the Armada's hard-pressed store of victuals. There was, however, no guarantee that local friendliness would last, even if the dangers of the anchorage and the exposure to enemy attack could be endured.

Notwithstanding its disadvantages, Medina Sidonia had some reason to hope that Calais would suffice as a place 'where Parma might join him'. He knew that Parma would not trust, unescorted, to the open sea. But to get from Dunkirk to Calais he would have only thirty miles of inshore sailing along a neutral coast, where for most of the way the shallow water would protect him from the English galleons. It now appears improbable that Parma would have risked even that: his letters during the critical period when the Armada was in reach all insist on his inability to emerge—by implication, until the Armada had beaten the English fleet. From Medina Sidonia's point of view, however, Parma's task seemed reasonable. Aboard the Armada, Parma's naval strength was evidently grossly overestimated: Medina Sidonia thought he could borrow thirty, forty, or even fifty flyboats from Dunkirk, whereas Parma's strength in pinnaces and flyboats was only forty sail all told, and, if he was to run the gauntlet of the Dutch, it would be he who would need to borrow armed pinnaces from the Armada, rather than the other way round. It was recognized, none the less, that Parma's sortie would be risky. He might lose half his men, Don Diego de Pimentel exclaimed, presumably in exaggeration, but at least he would surely try to make a move. This expectation—or, rather, hope—was never to be put to the test. Parma's embarkation was behind schedule—by six days, at his own estimate, or a fortnight by that of observers from the Armada. And in any case, the wind was unfavourable, and would remain so for the duration of the campaign.

The Armada could not wait for Parma to embark or the wind to change. 'It was impossible for it to remain where it was', as the duke warned Parma, and as everyone agreed, 'without very great risk.' The greatest risk was from those 'diabolical machines' and 'incendiary contrivances' which the English were expected to be able to unleash against the fleet. 'Have a care', the king's instructions had warned, 'to have the Armada well guarded and watched so that the enemy may not set fire to any of the ships, in which connexion there is great need to be wary, against people will stop at nothing and have plenty of devices for this purpose.' Shortly before the Armada sailed, Medina Sidonia was warned in a report from the royal secretary, Juan de Idiáquez, to beware of the exploding fireships and floating mines. These had been used with terrible effect at Antwerp and again, unsuccessfully, at Dunkirk by the Dutch, and their inventor, the Italian engineer Giambelli, was known to be working in England. In fact, not only did the English have none of these diabolical 'hellburners'—for Giambelli's job had been to build the valueless Thames barrage, which collapsed under its own weight—they had no fireships, either. This was a serious oversight, since, had English fears been realized and the Armada been used to seize an English port, fireships would have had an obvious role.[2]

They were equally obviously the best means by which to seek to dislodge the Armada from the Calais roads; yet when they were needed they could not be had and had to be extemporized from among the small vessels of the fleet. Their usefulness was apparent to both sides. Sir William Wynter, who was never much given to modesty, claimed that it was his idea to use them. 'Having viewed myself the great and hugeness of the Spanish army ... [I] did consider that it was not possible to remove them but by a device of firing of ships.' On the Spanish side, the stratagem was foreseen and precautions taken. What made the recourse to fireships—and even their timing—predictable was not only the vulnerability of the Armada's position to this type of attack, but also the fact that they would be sped on their way, if unleashed during the evening tide, by a north–south current of $\frac{3}{4}$ knot and a flood stream racing at $2\frac{3}{4}$ knots. Fireships were a common device and there were established procedures for dealing with them, even with this accession of speed, by catching them with a grappling cable from a screen of fast pinnaces and towing them away while they were still far off. This was a tricky operation, especially in a rapid current under long-range guns, but if it failed, or was only partly successful, the ships behind the screen could try to evade any fireships that got through and leave them to burn themselves out. The defensive tactics could only be countered by the attackers if the fireships could be launched in

Pine's version of the fireship attack presents some curious features. The fireships themselves are of duly modest proportions but are given pride of place in the centre of the composition. Two of them appear to be literally exploding, while the guns of the others are being fired by the heat. Unrealistically, however, their sails are furled. Notice the crews rowing back to the English fleet, and the Spanish pinnaces struggling to tow the fireships away. Neither main fleet is shown in good order, but the chaotic appearance of the Armada, in particular, contrasts with the precision of the half-moon formation depicted in Pine's other scenes (pp. 111 and 175).

The artist of the Armada playing-cards also gives prominence to the fireships, billowing smoke as they bear down on the Spaniards. He seems to have increased their number to nine. Note the prominent anchor, just emerging from the waves in the centre foreground.

overwhelming numbers, or at a fleet in a very cramped anchorage, or filled with explosives, like the 'hellburners', which would scatter flaming wreckage far and wide. None of these conditions was fulfilled by the English fireship attack on the Armada. The fireships were large—averaging nearly two hundred tons—but not numerous, having to be recruited from the fleet and prepared within a few hours. they were manned by a hundred volunteers, who were later paid five pounds between them for their service. There were only eight ships, at most, of which six penetrated the Spanish screen. The Calais roads, though hard to manœuvre in because of the currents, afforded ample space, and there was only one collision in the course of the panic that overtook the Spanish fleet that night. The fireships were rudimentary and without sophisticated devices and, in the event, caused no fires aboard the Spanish fleet. By all rational expectations, they should have failed.[3]

The fear that the fireships concealed explosive devices, however, caused Spanish morale to snap. At midnight Medina Sidonia saw two fires emerging from the English fleet 'and these two gradually increased to eight'. They approached ablaze, with rudders lashed and sails streaming from the yards 'and a fearsome fire at the prow, spreading towards the stern',

Hendrik Vroom's evocative painting of the fireship attack in Calais roads captures the moment at which the Armada slipped its cables and fled in confusion. In the background Calais is shown with the number, shapes, and distribution of its spires very much as in the work of Robert Adams. In the centre of the middle ground the collision. which disabled the leading galleass is shown. Most of the ships in the foreground are already making their escape, with sails unfurled; on some, raised anchors can be seen. Note the warning guns being fired and the water made turbid by a loose anchor-cable in the left-hand foreground.

towed by launches that cast off and fled as the *San Lorenzo*, which was alongside the flagship, opened fire. Perhaps because they were so large, perhaps because 'diabolical contrivances' had been so anxiously awaited, it was universally assumed on the Spanish side that they 'were some of the infernal machines'. Many Spaniards remained convinced that this was just what they were, even after they had burned themselves out harmlessly. Pedro Coco Calderón observed their failure to cause any damage, but still referred to them as explosive machines. A well-informed—though, of course, propagandistic—Spanish broadside claimed that it was only smart seamanship that saved the fleet from the incendiary devilry of the English:

My lord duke, foreseeing the danger, obviated it by ordering that the cables should be cut from the ships which lay nearest, and the anchors be raised from the others incredibly swiftly. And with this, the enemy being unable to prevent it as he intended, our ships sailed out elegantly and with such precision that, had it not been so, our Armada would have fared ill, for, in the very place we had just vacated, those fireships sent up a discharge of so many ingenious contrivances as would have been sufficient to set the whole sea alight.[4]

The reality of the Armada's attempt to evade the fireships seems to have been almost the opposite of the impression contrived by this account. In place of the implied discipline there was panic; in place of the precision, chaos. Few, if any, anchors were hoisted; almost all cables were cut; and because every ship had several kedges dropped against the current, 'a hundred thousand ducats' worth of anchors and cables', Fray Bernardo de Góngora lamented, 'were left on the bottom'. Recalde's ship left three anchors in the roads. Later in the voyage she had to swap fragments with the Castilian galleon *San Juan Menor* in order to patch up a Heath-Robinson means of holding fast to the bottom. The *Santa María de la Rosa* had only one anchor to drag on when it struck a reef and sank in the Blaskets. The *Regazona* and *Florencia* got back to Spain with no anchors at all. These instances could be repeated indefinitely throughout the fleet.

With daylight on the 8 August, the Armada could be seen to have scattered 'in a thousand directions' and a good many of the ships to have fled. It is not clear now—and perhaps was not clear at the time—what Medina Sidonia's orders for the night's manœuvre had been. Passed by word of mouth from ship to ship, hurried and probably inexplicit, they would have been intended by the commander to be understood in the context of routine methods of responding to attacks by fireship. The duke claimed to have issued orders to the fleet that 'when the fires had passed, they were to return to the same positions again'. In fact, except for the leading galleass *San Lorenzo*, which collided in the panic of the night and lost her rudder, all the ships that remained, including the flagship herself, edged a mile or two out to sea, but it was reasonable, and consistent with the usual drill, for the duke to expect them to keep together and maintain their overnight positions as nearly as possible. This was the first of two occasions when large numbers of ships broke or ignored the duke's orders with disastrous results. Does this vindicate Medina Sidonia's leadership, or condemn it? On the whole, the better part of leadership is to inspire confidence and command obedience: Medina Sidonia almost invariably made sound decisions; he must be said to have failed, however, to create the sort of relationship with his subordinate captains that would have inspired them to apply his orders with unfailing discipline. At Calais the fleet had lain, idle and apprehensive, in a dangerous and exposed position, surrounded by foes, cut off from hope of help, exasperated by Parma's apparent inertia, on a mission in which even the commander had no confidence and which was widely thought to be technically impossible. The Spaniards, in the recollection of Don Luis de Miranda, had spent the night 'resolved to wait, since there was nothing else to be done, and with a great presentiment of evil from that devilish people and their arts'. Nerves already jangled by a running firefight up the Channel were exposed to fears and rumours of explosive machines. In these circumstances, when those very machines seemed to burst into the anchorage, it is not surprising that individual captains should have wished to put as much water behind them as possible. Medina Sidonia, in his official report, characteristically refrained from blaming his subordinates: 'The current was so strong that it drove our Armada in such manner as, although the flagship and various ships that were near her anchored again, firing a piece of ordnance, the rest did not see her and so were driven off towards Dunkirk.'[5]

The English seem to have been surprised by their good fortune. It was only gradually, to all appearances, that realization grew on them that the Armada was no longer intact. Howard's first thought on the morning of the 8th was to make a prize of the great galleass which he could see wallowing rudderless in the mouth of Calais harbour, alone, except perhaps for the embers of the fireships. Aboard the *San Lorenzo*, Don Hugo de Moncada realized that he must reach the safety of the harbour if he could, and hope that the English would not dare to violate French neutrality. Alternatively, he could use the relatively shallow draught of the galleass to drive her aground where the English heavy ordnance could not reach. In the course of the pursuit, he settled for the latter course, but came aground rather high, and at an awkward angle, where he could not bring his guns into play. This gave the English some hope of reaching her with boarding-parties in boats, but for hours the crew of the *San Lorenzo* held them off with small-arms fire, while Hawkins stood to with a squadron at the harbour's mouth. These dramatic turns of fortune continued, first, when Don Hugo was felled by a musket shot and the other defenders, losing heart, began to scramble and swim for safety, while the attackers swarmed over the sides. Then, when it seemed as if the English had taken possession of the vessel, the garrison of Calais opened fire with heavy guns to drive them away, presumably to assert French rights to a wreck abandoned on French shores. By then Howard, thinking that the struggle for the galleass was over, had left to join the battle. He could not protest to the governor of Calais, he wrote to Walsingham, 'because I was in fight; therefore I pray you to write unto him, either to deliver her, or at leastwise to promise upon his honour that he will not yield her up again unto the enemy'.

English dilatoriness saved the Armada. By rapid pursuit the English could have isolated the fighting galleons, scooped up many scattered hulks as prizes, and driven the leemost Spaniards on to the shoals. First, however, Howard failed to recognize his chance; then he let his attention be absorbed by the chance of a rich prize at Calais. Meanwhile, Medina Sidonia was given time to assemble a rearguard behind which the Armada could reform or flee. His own account of the interlude before battle began in earnest ran thus:

At dawn on Monday, the 8th, the duke, seeing that his Armada was far ahead, and that the enemy was bearing down upon us with all sail, weighed his anchor to go and collect the Armada, and endeavour to bring it back to its previous position. The wind freshened from the north-west, which is on to the shore, and the English fleet of 136 sail, with the wind and tide in its favour, was overhauling us with great speed, whereupon the duke recognised that if he continued to bear room and tried to come up with the Armada all would be lost, as his Flemish pilots told him he was already very near the Dunkirk shoals. In order to save his ships he accordingly determined to face the whole of the enemy's fleet, sending pinnaces to advise the rest of the Armada to luff close, as they were running on to the Dunkirk shoals.

There is a note of self-praise here, conspicuous precisely because it contrasts with Medina Sidonia's usual humility. Coming from Medina Sidonia, it carries conviction. he aimed to 'save his ships' and 'face the enemy', not as a romantic hero, which he could never be, but as a good husband-man, which he was, whose duty was spare his king's resources. Only in trying to give the impression that the Armada was more collected and in better order than was really the case—and this, one suspects, more for his subordinates' sake than his own—does this passage of Medina Sidonia's account exceed reality. Some ships were close by him and others answered his call. The hulk of Pedro Coco Calderón was already on its way back towards its previous anchorage, in conformity with what was known or presumed of the admiral's orders. By the time she came up with the flagship, four other galleons were already there. It is not clear at what stage the battle began but it was probably while the *San Martín* was still alone or supported only by these ships. Gradually, however, more of the Spanish fighting ships joined the line. The sea was heavy—certainly heavier than any they had fought in during the run up the Channel; the wind and current were against them; they were joining an action already dominated by large numbers of handier and more heavily gunned enemy ships. Yet they were able to reconstruct a recognizable defensive formation and maintain it for much of the rest of the day. These facts establish two judgements beyond cavil: Spanish seamanship, which has often been blamed for the failure of the Armada, was by no means deficient in prowess, at least as far as the major fighting ships are concerned; and Spanish morale, though it had failed aboard many ships the night before, was strong, or at least quickly recoverable, in the most important part of the fleet. By the time Sir William Wynter joined the fight the Spaniards were in

the proportion of a half-moon. Their admiral and vice-admiral, they went in the midst, and the greatest number of them; and there went in each side, in the wings, their galleasses, armados of Portugal, and other great ships, in the whole to the number of sixteen in a wing, which did seem to be of their principal shipping.

Evidently, the Spaniards were trying to rebuild the defensive carapace that had protected their fleet in the Channel.[6]

The English attack was led by Drake's squadron. For a reason which remains obscure, however, Drake contributed only a short exchange of fire to the battle, perhaps with Medina Sidonia and Oquendo, before withdrawing. Frobisher accused him of cowardice or knavery—'I know not which, but the one I will swear'. The *San Martín* still had a good supply of shot and, as we have seen, was capable of outgunning the English at medium range, and it is therefore possible that Drake was acting with discretion rather than valour. Alternatively, as Frobisher seems to have thought, he may have preferred to go off seeking prizes rather than become embroiled in a long gun fight with ships too strong to be captured save by chance. The most likely explanation, however, is that he realized what Medina Sidonia was up to and wished to elude the Spanish rearguard while he could. If so, he cannot be said to have followed

Apparently by a seventeenth-century Dutch hand, this popular image of an Armada battle is highly romantic. Notice the queen, watching with her troops from an ideally picturesque English shore somewhat too well equipped with impressive fortifications. The galleons and fighting hulks, however, are faithfully depicted.

up his insight very effectively: if he chased any of the rest of the Spanish fleet, he failed to catch it, much less to seize or sink it. Drake's part did not go much beyond the sparring already practised during the Armada's voyage up the Channel, and the initial impression made on some observers was that the battle of 8 August was yet another encounter of the same sort, in which no tactical innovations were tried.

The day after the battle, Robert Cecil sent news to his father of 'a proper gentleman of Salamanca' who had been captured at Calais, and of 'that fight which we saw upon the land yesterday; where, terrible as it was in appearance, there was few men hurt with any shot, nor any one vessel sunk. For, as this man reporteth, they shoot very far off, and for boarding our men have not any reason.' Cecil's appraisal, however, was misleading for any but the very earliest stages of the fight. We are well informed about two sectors of the battle—in the centre of the Spanish line, around the flagship, and on the weather wing, against the great East Indiaman the *San Felipe*—where for several hours the fighting was within small-arms range. Almost all the sources concur in Henry Whyte's summing-up: 'As soon as we that pursued the fleet came up within musket shot of them, the fight began very hotly.' Around the *San Martín*, this must have happened fairly early in the day; Medina Sidonia's account must exaggerate when it speaks of the assault lasting non-stop from daybreak, but may be correct in associating the beginning of the close fight with the arrival of Frobisher's *Triumph*—called, as usual, 'the enemy's flagship' in the Spanish account, because it was the biggest vessel in the English fleet: 'supported by most of his fleet, he attacked our flagship with great fury at daybreak, approaching within musket-shot and sometimes within half-musket-shot'. The attack lasted until three in the afternoon, without a moment's cessation of the artillery fire. Fray Bernardo de Góngora thought the *San Martín* had thirty ships against her. Apart from those which remained at Calais to try to take the *San Lorenzo*, 'all the other enemy ships fired on us so fiercely that it was a trial.' Observers in Calais thought they were witnessing 'firing on both sides that was the greatest that was ever seen or imagined'. Pedro Coco Calderón, whose hulk was close to the flagship and to the *San Marcos* for most of the duration of the battle, gives the most circumstantial account (though his recollections, jotted down in more or less random order, have to be carefully reshuffled to make chronological sense):

The enemy attacked our flagship with a great fusillade from seven in the morning, which lasted for more than nine hours, and on the starboard side fired so many shot that more than two hundred struck the sails and the flank of the hull, killing and wounding many men; and they caused the loss of three great guns, knocking them from their carriages so that they could no longer serve, and they tore through much of the rigging, and from the shot which pierced the waterline, the galleon was taking in so much sea that two divers could hardly mend the leaks, working at it with tow and lead plates and working both pumps all day and night. The men were exhausted from the many tasks they had faced during the previous night and from helping to manhandle the guns without having been given anything to eat.

Aboard the flagship, Fray Bernardo confirmed the loss of personnel in slightly different terms; 'some of our men died, but none of quality, though it was a miracle that the duke himself escaped'. Pedro Coco attributed a prominent role to his own hulk:

We gave them our luff as much as we could … giving them our prow and broadside and bringing half our poop guns into play, enduring the hail of shot and returning it … Some of our men were killed and our hull, sails and rigging badly mauled, so that we had to change the mainsail, and we shipped a lot of water where she was pierced by shot.

The intensity of the battle in the centre of the line is illustrated by the fate of other ships, too. The *San Marcos*, which, Medina Sidonia reported, stood by him throughout, the *San Juan Bautista* of Diego Flores's squadron, the Castilian galleon, *San Juan Menor*, under Don Diego Enríquez, which now seems to have led the Andalusian squadron, the *Rata Encoronada* of Don Alonso de Leiva, Oquendo's galleon and Bertendona's carrack were all singled out in the leading accounts for their valour displayed and damage sustained, 'unfit', in the duke's opinion 'to cope with further attack', at least until the following day. Oquendo reported, for instance, that he was 'frequently hit and badly mauled and the foremast and mainmast have two balls of great shot imbedded in them'. Eventually, however, the tenacity of resistance wore down the English attack, which, in this quarter, seems to have subsided at about the third or fourth hour of the afternoon.[7]

By then, however, the battle was building up to an even more ferocious level of intensity on the weather wing of the Spanish line, where Seymour's ships and later, perhaps, Howard's own squadron had concentrated their attack. Here, Pedro Coco Calderón has left a vivid description of the fiercest and most destructive passage of the battle, on the leemost tip of the Spanish line, held by Don Francisco de Toledo's *San Felipe*. With help from Medina Sidonia's account, the fight on this wing can be reconstructed in some detail. The narratives begin in earnest at the point when the *San Felipe* ran out of great shot. Until then, she had maintained her position at the end of the line, apparently suffering only the sort of superficial damage typical of the Channel fights. When her guns fell silent, however, the English crowded round to pound her at close range: sixteen ships surrounded her, according to Pedro Coco Calderón. They would still not close to grapple but came to within arquebus range, pouring in shot which took away the rudder, the mainmast above the mainyard, and most of the rigging, and wreaked terrible slaughter on the deck. She was holed in the course of the day, but probably not yet: damage to hulls was cumulative, and divers and carpenters worked quickly to patch any place that was pierced near the waterline so that it is not surprising that it took several hours of close bombardment to bring a ship near to sinking. Medina Sidonia's laconic summing-up says only that 'the enemy turned upon him with so hot an artillery fire that he was in difficulty'. The *San Mateo* of Don Diego de Pimentel, probably the next ship in the line, was the first to try to relieve the *San Felipe*, but in her turn was surrounded by a large number of English ships—ten, says Pedro Coco. Again she was hounded very closely and must almost have managed to grapple, if we can believe Pedro Coco's story of the Englishman who succeeded in leaping aboard, only to be cut to pieces by the defenders. More ships joined the rescue attempt, including the *Santa María de Begoña*, the *Trinidad Valencera*, and the Ragusan *San Juan de Sicilia*, the latter two both Levanters and therefore highly vulnerable to heavy ordnance at close range. This must have been the encounter in which the Ragusan was 'so battered by cannon-balls that she could move forward with great difficulty . . . as best she could with her sails in tatters, her rigging torn and her masts broken', according to one of her Croat crewmen who got back to El Ferrol. Pedro Coco confirmed that all her sails had to be replaced and that the ship 'showed great strength and courage'. The Italian vessel of which some

Dutch deserters claimed they caught a glimpse 'through the portholes, all full of blood, which yet maintained the fight in her rank three hours after' may have been the *Trinidad Valencera*, which was assigned an honourable place in Medina Sidonia's relation. According to Medina Sidonia, these ships 'very nearly closed with the enemy without grappling, the English keeping up their artillery fire, from which our men defended themselves with musketry and arquebus fire, as they were so near', perhaps because the shortage of shot was now general among the engaged ships, or perhaps because the Spanish force was practising the classic boarding tactics, approaching bowsprit-on. English accounts also record some moments in the ensuing mêlée. Seymour, 'with the *Vanguard*, the *Antelope* and others, charged upon the tail, being somewhat broken, and distressed three of his ships'—the reference is probably to the *San Juan de Sicilia*, *San Mateo*, and *San Felipe*, which were hindmost for most of the time—'among which my ship shot one of them through six times, being within less than musket shot'. This was probably also the juncture of the battle described by Wynter in a memorable passage of his letter to Walsingham, claiming that he withheld his fire until within sixty yards of the enemy and exclaiming, 'I deliver unto your Honour upon the credit of a poor gentleman, that out of my ship there was shot 500 shot of demi-culverin; and when I was furthest off in discharging any of the pieces I was not out of shot of their arquebus'.[8]

It was still the *San Felipe* and the *San Mateo*, now both apparently out of great shot, that continued to endure the worst pounding. Five of the *San Felipe*'s starboard guns had been disabled by English gunfire—showing that she had been giving her luff during the worst of the fight, heedless of the shoals. A sixth was spiked by a gunner in apparent despair. The uppermost deck was all shot away. Both pumps were destroyed. There was no rigging left. Don Francisco's response was to get out the grappling hooks and challenge the foe to hand-to-hand combat. By now the enemy were so close that at least one ship was within hailing distance—perhaps Wynter's, as he later claimed that he was 'many times within speech of one another'. Don Francisco's challenge was ignored, but an Englishman called from the mainyard to invite them to surrender, 'Good soldiers that ye are, surrender to the honourable terms which we will grant you.' Pedro Coco Calderón's account continues:

And a musketeer, instead of giving him answer, felled him with a shot in full view of everyone, and after that the general ordered all the muskets and arquebuses to fire, which, when the enemy saw it, made them draw off, with our men calling them cowards and declaring their lack of courage in insulting terms, calling them chickens and Lutherans and that they should re-join the battle.

The reason for believing Pedro Coco's account is that his ship formed part of a task force, with the flagship, sent from the middle of the Spanish order to mount a last attempt to rescue the *San Mateo* and the *San Felipe*. He was probably therefore close to the action by now. Medina Sidonia believed it was his intervention that drove the English away:

The duke heard the sound of small arms, but was unable to distinguish what was going on from the maintop, in consequence of the smoke, but he saw that two of our ships were amongst the enemy, and that the latter, leaving our flagship, concentrated all his fleet in that direction. So the duke ordered the flagship to put about to assist them. The duke's ship was so much damaged with great shot between wind and water that the inflow could not be stopped, and her rigging was almost cut to shreds, but when the enemy saw that she was approaching, his ships left the vessels they were attacking, namely those of Don Alonso de Luzón [*Trinidad Valencera*], Garibay [*Santa María de Begoña*], Don Francisco de Toledo [*San Felipe*], Don Diego Pimentel [*San Mateo*], and Don Diego Tellez Enríquez [*San Juan de Sicilia*].

It is more likely, however, that it was worsening weather and the apparently inexorable drift of the contending ships on to the Flanders shoals that made the English break off the fight. If Wynter's experience was typical, moreover, the English too were now running short of great shot. By seven o'clock the Armada's rearguard was at peace to nurse its wounds and try to correct its course. The *San Mateo* posed the most serious immediate problems: she was far to leeward, manifestly unserviceable, and falling away beyond reach of help. Medina Sidonia sent a pinnace to take off the crew, but Don Diego Pimentel refused to leave his ship, requesting only a pilot and diver to help him save her, though it seems that they never reached her as she drifted too far into danger meanwhile. Don Francisco de Toledo and his ship's captain, Juan Pozas de Santiso, were victims of a similar heroic gesture aboard the *San Felipe*. At seven o'clock they fired a signal for help, as they believed they were on the point of sinking. The hulk *Doncella* reached them and took the entire company off; but, in Pedro Coco's version of the story, Juan Pozas believed the hulk was likely to go down too: 'If I must die here,' exclaimed Don Francisco, 'I had better do so in my own galleon,' and they both sprang back on to the evacuated ship, 'which was a great misfortune,' as Medina Sidonia added, 'for it was not true that the hulk was sinking, and the *San Felipe* also went towards Zeeland with Don Francisco on board'. Pedro Coco saw the *San Felipe* and *San Mateo* drifting off to their doom together, far off, so close, poop to poop, that he thought the stauncher ship had pinioned the other to keep her afloat.

Evidently English tactics in the battle of 8 August represented a considerable advance on those of the days of sparring in the Channel. For the first time, frequently and protractedly, if selectively, they had closed to within short range, especially in their duel with the Spanish flagship and in the bloody contest on the weather wing of the Spanish line. Still, 'for boarding they had no reason'. Indeed, so determined were they in their continued avoidance of it—save in the case of the excessive zealot who was said to have leapt aboard the *San Mateo*—that the Spaniards suspected, quite falsely, that their enemy must have 'an order from the queen, under pain of death, that by no manner of means should they board any ship of our Armada'. While continuing to rely exclusively on fire-power, however, they took new steps to make it effective by plying their heavy ordnance at ship-smashing range. Wynter's alleged five hundred rounds of demi-culverin—officially a long-range type of ordinance—at less than sixty yards demonstrated vividly English determination to learn from and break with the experience of the previous few days.

The result, however, was far from satisfactory from an English point of view. By the most significant standard—that of objectives fulfilled or disappointed—the battle of 8 August, so often acclaimed as an English triumph and even as one of the decisive episodes in the history of sea power, was a Spanish victory and an English defeat. The Spaniards had achieved their purpose: the Armada was saved. The English had failed in theirs: it escaped, scotched but not killed. Howard thought that he had only 'plucked its feathers'. The Spaniards, though apprehensive about how long their fleet could remain battle-worthy, were willing to renew the fight next day.

Against the background of the sparring in the Channel, what seems most remarkable about the battle of 8 August is the ineffectiveness even of the modified English tactics, and the feebleness of even their close-range gunnery. The *San Martín*, despite the ferocity of the fusillade to which she was subjected, by all accounts, for several hours, emerged bloodied but unbowed. She seems to have been much the most severely mauled of the ships in the centre of the Armada's line, closely followed, perhaps, by the *San Marcos* and the *Florencia*. Even where the fight was closest and most formidable, on the weather wing of the Spanish line, the only disabled ships were the two which ran out of shot at a critical stage, the *San Mateo* and *San Felipe*. The *San Juan de Sicilia* was almost destroyed, but managed to limp away under the protection of the other great ships, 'making shift to follow in very bad case'. She went on to survive all the subsequent bad weather. Only one ship sank outright as a result of English action—the Guipúzcoan armed merchantman *María Juan*, which became detached from the fleet and surrounded. The facts seem to endorse the judgement of George Birmingham's parson: 'those Spanish galleons were pretty tough ships'. English ships—to judge from the fact that they all emerged, if not unscathed, without fundamental damage—seem to have been even tougher.[9]

The Spanish response to English gunnery was even feebler and, indeed, apparently negligible in its effects. Many of the reasons for the failure of the Spanish guns in the Channel fights were still impairing their effectiveness on the 8 August. But why, with the closing of the range, did the Spanish artillery not begin to tell more heavily? There were various extenuating circumstances, peculiar to that day. In the first place, as so often in the history of the Armada, the weather was against the Spaniards. The battle was fought in exceptionally heavy seas—heavy by the usual standards of battle conditions, particularly towards the end of the action, when a sudden squall blew up and impeded the work of the divers and the attempted rescue of the *San Mateo*. This squall was bad enough to finish off the *María Juan*, with the rescue of only a single boatload of survivors. Poor weather may also explain the small part played by the three galleasses, which were in place in the Spanish line and, in the early part of the action, viewed from Calais, apparently 'doing great execution', but which are not mentioned in any accounts of the fighting later in the day: they were, after all, reputedly, 'frail barks for rough seas'. The rest of the Spanish ships—even the Levanters and fighting carracks and hulks—seem to have manœuvred freely and, as Wynter acknowledged, 'in very good order', despite the weather; but what was little worse than normal to a seasoned mariner or naval gunner, was unaccustomed and challenging to the landsmen who serviced the Spanish guns. Secondly, some of the Spanish guns were put out of action by English fire: we hear of five guns disabled aboard the *San Felipe*, for instance, and three on the *San Martín*. Spanish guns were mounted higher than those of the English—as high as possible, Diego García tells us, in order to allow the gunsmoke to disperse quickly—and were probably much more exposed to this sort of misadventure. Thirdly, it seems that for much of the battle, precisely because of the gunsmoke, the Spaniards could not see what they were doing. Diego García tells us that one of the ancillary advantages of holding the weather-gauge in battle is that the wind blows the smoke on to the enemy ships. In the battle of 8 August, the Spaniards were on the receiving end. The smoke must have been particularly dense because of the exceptionally close range and exceptionally heavy rate of fire. The English, moreover, were of course using their notoriously poor-quality black powder and therefore generating peculiarly noxious fumes. When Medina Sidonia climbed his own rigging to see what was happening on

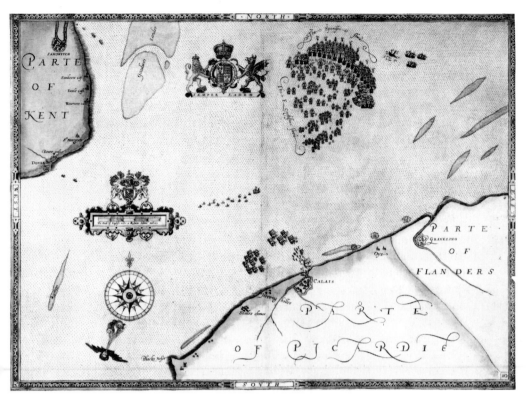

The close dependence of Pine's source on the work of Robert Adams is demonstrated by their views of the battle of 8 August. Both give great prominence to the struggle for the stricken galleass under the walls of Calais and show the French shore batteries in action. Both show the distribution of English ships standing off or joining the fray in much the same way. In both the *María Juan* is seen sinking and there are three Spanish ships dangerously to leeward towards the shoals. Adams's version is remarkable for the good order with which the Spanish battle-line appears to have been reconstructed, with the *San Martín*, shown larger than other vessels to reflect its importance, in the centre of the line.

his weather wing, he could not see for the smoke, although he heard the sound of small-arms fire. According to Fray Bernardo de Góngora, 'the confusion was the worst there has ever been in the world because of all the firing and smoke'. According to members of the crew of the galleass *Zúñiga*, interviewed in Le Havre on their way home to Spain, 'many of our ships were quite hidden by the smoke'. Furthermore, if Fray Bernardo is to be believed, the Spaniards were so busy struggling to keep off the lee shore that they had little opportunity for gunnery: 'all that day we were hauling on the bow-line against the wind so as not to be grounded on the banks and so our ships could not bring their guns into play as they wished'. But these and all the other reasons for the failures of Spanish gunnery mentioned in the last chapter seem to have been outweighed on 8 August by sheer shortage of shot. With the two Spanish galleons most closely engaged, the *San Felipe* and the *San Mateo*, the English do not seem to have closed until their great shot was exhausted and they could reply only with small arms. Fray Bernardo says that their gunners complained of starting the campaign with only thirty rounds apiece. According to Medina Sidonia, most front-line ships ended the battle with not another round on board. Don Francisco de Bobadilla attributed the failure of the Armada to insufficiency of shot 'on the fifth day' of battle. It may be that except for the *San Martín*, the only Spanish ships that had an opportunity to fight at ship-smashing range were those which had no shot with which to do it.[10]

Medina Sidonia's own summary of the battle honours of 8 August seems scrupulously fair: his ships were 'distressed and battered'; 'the shot had run out'; 'the queen's fleet was superior in this sort of fighting' because of heavier guns and handier ships and 'that of Your Majesty had the advantage only in small arms, and since we could not come to handstroke, experience has shown that this could avail us little'. A less modest man, taking a broader view, might have claimed a resounding victory. By saving the Armada from destruction, the Spaniards' *ad hoc* rearguard had achieved a notable success, against adverse conditions and superior forces. The English, even in the most favourable circumstances, had been unable to disperse or destroy the enemy or eliminate the threat of invasion. By the end of the day's fighting on 8 August, the Armada, since entering the Channel with 125 sail, had lost six ships, only two of which had been disabled by enemy action. In the circumstances, this was an acceptable rate of attrition. The outcome of the Armada's last battle, had that been the end of the story, would probably have been enough to secure Spain's strategic objectives, forcing the English into a peace that would involve concessions to Spanish terms and leave the prestige of Philip II intact.[11]

Eight

The End of the Armada

WHAT the English could not achieve in the battle of the 8 August, they may have hoped to see accomplished the following day by the wind and current and the menacing topography of the shore. It had been suspected in Calais that the English plan all along had been to usher the Armada onto the Flanders shoals. 'The Spanish fleet is powerful, only it has no port of refuge in these parts, and with bad weather may be driven onto the banks, which is the English plan, without thinking of coming to close quarters.' In the event, English tactics did seem to conform to that plan. They harried the Armada closely, it is true, but without coming to handstroke, and broke off only when the banks were perilously near.[1]

It must be acknowledged, however, that the chief danger to the Armada arose not because of English action but in spite of it. To judge from Drake's assessment of the battle of 8 August, the English purpose in hounding the Armada eastwards was not to drive it onto the banks but to prevent, or at least postpone, any junction with Parma. 'God hath given us so good a day,' he wrote at its close, 'in forcing the enemy so far to leeward as I hope in God the Prince of Parma and the Duke of Sidonia shall not shake hands this few days; and whensoever they shall meet, I believe neither of them will greatly rejoice of this day's service.' When the English withdrew from the fight in the centre of the Spanish line, in the middle of the afternoon, the Spanish ships in that quarter of the battle do not seem to have been in danger and were moving away from the shoals, according to Wynter, 'in very good order'. The ships on the weather wing must have seemed in greater peril when the English desisted at about six o'clock, but only the *San Mateo* and *San Felipe* proved to be too heavily disabled to recover their position and even they do not seem to have been definitively lost until driven farther to leeward by the late squall that sank the *María Juan*. Medina Sidonia seems consciously to have waived his chance to get right away to safety during the night: his honourable excuses were that he hoped to be in a position to return to the Channel the next day and that he still wished to protect his leemost ships from enemy attack.

Even on the morning of 9 August, despite a strong adverse wind from two o'clock to daybreak, it seemed at first as if the fleet would clear the banks. Medina Sidonia feared—perhaps falsely as there is no direct corroboration in English sources—that the enemy were about to press it further in an attempt to edge it into disaster. Further English intervention, however, soon began to seem unnecessary; the wind remained remorselessly in the north-west, strong enough to drive the Spaniards on. This was convenient for the English, but unforeseeable when they broke off the action on the previous night. Medina Sidonia's account is clear enough:

At two o'clock in the morning the wind blew so strongly that, although our flagship was brought up as close to the wind as possible, she began to fall off to leeward towards the Zealand coast, the duke's intention having been to stay, so that he might again enter the Channel. At daybreak the north-west wind fell somewhat, and we descried the enemy's fleet of 109 ships rather over half a league astern of us. Our flagship remained in the rear with Juan Martínez de Recalde, Don Alonso de Leiva, the galleasses, the galleon *San Marcos* and the *San Juan* of Diego Flores, the rest of the Armada being distant and a great deal to leeward. The enemy's ships bore down on our flagship, which came round to the wind and lay to; the galleasses placed themselves in front, and the rest of our rearguard stood by to repel attack, whereupon the enemy retired. The duke then fired two guns to collect the Armada and sent a pilot in a pinnace to order the ships to keep their heads close to the wind, as they were almost on the Zealand shoals. This prevented the enemy from approaching closer to us, as they saw that our Armada was going to be lost. Indeed, the experienced pilots who accompanied the duke assured him at this time that it was impossible to save a single ship of the Armada, as they must inevitably be driven by the north-west wind onto the banks of Zealand. God alone could rescue them.

With the sounding from Pedro Coco Calderón's ship at seven fathoms, the duke is said to have hailed Oquendo, who was alongside, 'Señor Oquendo, what shall we do, for we are lost?' 'Let Diego Flores answer that,' Oquendo replied. 'As for me, I am going to fight and die like a man. Send me a supply of shot.' Medina Sidonia, of course, had already asked Diego Flores the same question: it must have been at this juncture, aboard the *San Martín*, in conference with Don Francisco de Bobadilla, that an exchange reliably reported in a piece of later hearsay took place; they discussed 'whether they should surrender to the enemy because of some certain natural peril, to which Diego Flores only answered that it was not yet time'. Don Francisco thought even this guarded response was cowardly. The choice of shipwreck or surrender remained unresolved. 'From this desperate peril,' the official account continued, 'we were saved by the wind shifting by God's mercy to the west-south-west, and the Armada was then able to steer a northerly course without danger to any of the ships.'[2]

For once the weather had favoured the Armada. First, the north-westerly had helped to deter the English from harrying too closely for fear of the lee shore; then, before the Spaniards could come to grief, a 'Catholic wind' had arisen from the south to save them as if by a miracle. The outcome of the battle of 8 August was confirmed. The Armada was essentially intact and effectively undefeated.

Both sides were in some perplexity about what more, if anything, the Armada could or should now try to do in northern waters. On the English side, the chief fears were, first, that the Armada would force another battle on the queen's navy by trying to seize an anchorage on the east coast; or, if the wind changed, that it might return to the Channel, perhaps after repairing battle damage in some neutral port, and renew the fight; or that the Armada would yet provide a diversion that would allow Parma to slip across the Channel. None of these prospects could be contemplated with equanimity. In the immediate aftermath of the Armada's escape, the English, with their 'powder and shot well near all spent', and a diminishing supply of munitions available on shore, expected apprehensively to be called on again. Hawkins was convinced the enemy would try for an English port. 'I doubt not, with God's favour, but we shall impeach their landing,' he wrote to Walsingham. 'There must be order for victual and money, powder and shot, to be sent after us.' Fenner confirmed that it was want of powder, shot, and victuals that made the English incapable of renewing the battle, and the council of war aboard the English fleet was careful to set down in writing the same reasons for their decision to confine their operations to a limited pursuit of the Armada.[3]

Wariness of Parma, meanwhile, was assuaged by promptly returning Seymour and Wynter to their watch over the straits. In the evening of Tuesday, the 9 August, Seymour 'was commanded by my Lord Admiral, with Your Majesty's fleet under my charge, to return back for the defence of Your Majesty's coasts, if anything be attempted by the duke of Parma, and therein [I] have obeyed his Lordship much against my will, expecting Your Majesty's further pleasure.' The same order was conveyed to Sir William Wynter, who greeted it with even greater reluctance, deferring his compliance until he had taken every step to verify the authenticity of the command and to ensure that Seymour, as his immediate superior, concurred. However, the letter, in which Wynter records these facts also shows that many English ships thought their service was over. When Seymour's fleet stole away in the twilight, 'we had much ado with the staying of many ships that would have returned with us besides our own'.

The producer of the Armada playing-cards has not overlooked the English fleet's forgotten squadron, patrolling the narrow seas under Seymour until 6 August, and obliged to return reluctantly to its post on the ninth.

The Ld Hen: Seymor wth 40 Engliſh and Dutch Ships keeping the Coaſt of the Nether=lands to hinder y̆ Prince of Parma's coming forth.

The snake had been scotched, not killed. But the common sense of ordinary seamen suggested that that was enough.[4]

Aboard the Armada, only two possible courses of action seem to have been seriously considered. On the afternoon of the 9th, while Howard on the Ark Royal was holding the council that sent Seymour back to the straits, Medina Sidonia summoned an extended council of war to the *San Martín*. By his own account, the duke expounded the fleet's predicament in a way heavily loaded in favour of an immediate return to Spain:

The duke submitted the state of the Armada and the lack of great shot, a fresh supply of which had been requested by all the principal ships, and asked the opinion of those present as to whether it would be best to return to the Channel, or sail home to Spain by the North Sea, the Duke of Parma not having sent advice that he would be able to come out promptly.

Medina Sidonia does not seem to have given the council the option of considering an attack on an English port to give the Armada an anchorage on the east coast of England. The weakness of the fleet in terms of battle damage and depleted supplies of shot can be presumed to have ruled that out. Nor did he seriously weigh the other possibility, suspected, in particular, by Drake, that he would seek the refuge of a neutral port in Norway or Denmark in order to make repairs before facing the English again. As far as the duke was concerned, they were 'enemy lands'. The Spaniards, can hardly have felt sanguine about their chances of acquiring victuals or, more particularly, munitions on the eastern shores of the North Sea. The choice was between attempting to fulfil the king's orders at the risk of the fleet; or settling for the real if limited achievement the Armada had registered so far, while conserving its ships and soldiers for another day.[5]

Though it is hard to resist the impression that he was seeking to foist upon Parma blame for a decision he would have taken anyway. Medina Sidonia evidently saw no point in returning to the Channel, even had the wind served, as Parma's unpreparedness made the accomplishment of the fleet's primary mission impossible. Nevertheless, the king's orders could not lightly be set aside. The royal watchdog aboard the *San Martín*, the duke's naval adviser Diego Flores, spoke in favour of a return to Calais: that, at least, was the hearsay reliably reported by Pedro Coco Calderón. So, according to Vanegas, did Recalde. Medina Sidonia merely recorded that 'the Council decided unanimously in favour of returning to the Channel if the weather would allow it, but, if not, then that they should obey the wind and sail to Spain by the North Sea, bearing in mind that the Armada was lacking all necessary things, and that the ships, which had resisted hitherto, were badly distressed'. Since, at the time, the Armada was flying north on the back of a gale, the problem must have seemed to have small practical significance. The captains were uttering a dutiful disclaimer in the direction of the king, while bowing to the demands of their circumstances. The English, in similar need, and equally anxious to indemnify themselves against accusations of dereliction of duty, took a closely comparable decision in a council of war of their own the day after next:

We whose names are hereunder written have determined and agreed in Council to follow and pursue the Spanish fleet until we have cleared our own coast and brought the Firth [of Forth] west of us; and then to return back again, as well to revictual our ships, which stand in extreme scarcity, as also to guard and defend our coast at home; with further protestation that, if our wants of victuals and munition were supplied, we would pursue them to the furthest that they durst have gone.

The memorandum was signed by an irreproachable combination of aristocratic propriety and nautical experience: Howard, Lord Thomas Howard, and Lord Sheffield; the Earl of Cumberland, who served with his own ship, much practised in piracy, as a gentleman-adventurer; and Drake, Hawkins, and Fenner. Edward Hoby, who countersigned the document, was the queen's watchdog aboard the English fleet. He had the watching brief, but not the advisory role, of a Diego Flores de Valdés. He was a clever lawyer, a university man, and a kinsman and client of Burghley's. He followed up his experience of the Armada by translating Bernardino de Mendoza's book on the theory of warfare into English. His adherence to the decision of the council of war guaranteed the professionals' invulnerability. As Hoby had kept watch for the queen over the preparations for the campaign, his signature also tended to validate the professionals' implied indictment of royal economies. The Spaniards were sailing away 'in obedience to the wind'. The English declined to pursue in obedience to equally imperious circumstances of their own making: shortage of the means to prosecute their advantage to victory.[6]

As the Armada entered the North Sea, morale, frayed by the experience of Calais, remained fragile. The secret report later drawn up for the king in Spain by Don Juan de Cardona on 'who did their duty in the recent expedition and those who did not' gives the impression that the fleet was in the throes of a crisis of discipline. Although in the battle of the 8 August, 'I understand that the majority did their duty', Cardona found that 'after they left the Channel all the ships thought of nothing but to steer their course and get back to Spain'. On the 10th at least two ships appeared to disobey orders and bolt for safety when the Armada was ordered to strike sails and offer battle again to the pursuing enemy. On the 11th, Medina Sidonia acted to stop the rot. Twenty captains were arraigned, or summarily condemned, for cowardice—'for having behaved badly and for being cowardly', in the words of Alonso de Vanegas; in some quarters of the fleet—in Pedro Coco's account, for instance—their offence was thought to have been committed in the battle of the 8th. In the only two cases of which we have a detailed report, however, it was for breaking ranks on the 10th that the captains were accused. One of them, Francisco de Cuellar, of the Castilian galleon *San Pedro*, has left a vivid account.

The galleon *San Pedro*, in which I shipped, was badly distressed by many rounds of great shot which the enemy lodged in her in various places, and although these were repaired as well as possible at the time, there were some inaccessible holes pierced by shot through which much water entered. After the fierce battle we had off Calais on 8th August, continuing from morning until seven o'clock in the evening, which was our last day's fight, while our Armada was withdrawing—ah, it grieves me to recall it!—the enemy fleet pursued to harry us from their country. When this was over and the danger had passed, which was

Howard indemnified himself against charges of cowardice or treason by justifying his decision to break off the pursuit of the Armada in this document, adopted by his council of war and countersigned by the queen's representative Sir Edward Hoby.

at the tenth of the month, we saw the enemy stand to, so some ships of our Armada furled sails and repaired their damage.

On this day, I, for my most grievous sins, was resting a little as for ten days I had not slept nor ceased to assist wherever I was wanted. Without saying anything to me, a wicked pilot I had with me made sail and passed out ahead of the flagship for about a league, as other ships had done, in order to make repairs. We were about to lower sail to see where the galleon was leaking when a pinnace came alongside and ordered me in the name of the duke to go aboard the flagship. I hastened to her, but before I could get there, orders were issued to another ship that I and another gentleman, Don Cristóbal de Ávila, who was sailing in command of a hulk which was way ahead of my galleon, should be executed in the most ignoble manner.

When I heard of this harsh sentence, I thought I would explode with indignation, calling on all to bear witness to the great injustice inflicted on me, who had served so well, as written depositions would show. The duke knew nothing of this because, as I should explain, he had retired and would see no one. It was Don Francisco de Bobadilla alone who issued and cancelled orders in the Armada and by him and others, whose wicked deeds are known, was everything regulated. He ordered me to be taken to the fiscal's ship for his sentence to be carried out. There I repaired, and though he was a severe judge, the fiscal heard my case and took testimony concerning me. He heard that I had served His Majesty as a good soldier and therefore became unwilling to execute the orders he had received. He wrote to the duke about it, saying that unless he received a direct order written by the duke and signed with his own hand, he would not comply with his orders, because he saw that I was not to blame, nor had I been responsible for the misdemeanour. I wrote a letter of my own to the duke to go with it, of such a kind that it made him re-consider carefully, and he wrote to the fiscal that he should not carry out the order against me, but execute that against Don Cristóbal, whom they hanged with great cruelty and dishonour, for he was a gentleman of renown.

In all, twenty culprits were condemned to death, but all save Don Cristóbal, who commanded the well-equipped hulk *Santa Barbara*, were spared. The luckless corpse was paraded through the fleet, dangling from the masthead of a pinnace, *pour encourager les autres.*[7]

The Armada's long periplus home was not adopted in a consciousness of defeat or a spirit of flight. It was simply, in the circumstances, the most rational choice. In the course of that voyage, had the direction of the wind and the severity of the weather been within the limits suggested by experience, or even had they been only somewhat worse than usual instead of attaining and exceeding unprecedented perversity and savagery, the Armada would have returned home, almost intact, with few further losses, having demonstrated the extraordinary reach of Spanish seapower and the vulnerability of the English in their native seas. The fleet's home-

coming would have been hailed as a triumph and their battles against
the English would have been recorded in the annals of war as significant,
if indecisive, victories. Pedro Coco Calderón later claimed to have uttered
dark forebodings at the time of the decision to continue north:

> The council decided on a return to Spain and when Captain Alonso de Benavides
> and Captain Vasco de Carbajal asked him what the route was like, he replied that
> there would be no want of intolerable hardship, because to get back to Corunna
> we would have to pass by England, Scotland, Ireland and the Isles about them,
> across seven hundred and fifty leagues of wild seas, little known to us. And then
> he made a test of samples of the biscuit and water he had on board, for everything
> else was in short supply, and more particularly on his hulk than in the rest of
> the fleet.

It was easy to write like that with hindsight, however; at the time, there
can have been little cause for apprehension. It could be foreseen that there
would be a long period of short rations, as a result of spoiling, but it is as
well to remember that the English had been on short rations since before
the campaign began and that the Armada remained, relatively speaking,
a well-provisioned fleet. The voyage would be protracted, as everyone of
experience knew, by the need to make a wide sweep into the Atlantic on
the way home, borne—as was correctly foreseen—by the prevailing north-
easterlies encountered to the north of Scotland, both to avoid the treacherous Irish coast and to cope with the prevailing westerlies in Irish
latitudes. It could reasonably be hoped that the season would be the
warmest and one of the least unsettled in northern seas, with the voyage
completed, or sufficiently advanced, by the time of the worst of the
September gales. While the particular route envisaged was naturally
unfrequented by Spanish shipping, it is nonsense to say that the seamen
of the Armada were unfamiliar with northern waters. Spaniards and
Portuguese had taken part in the Iceland and Baltic trades and fished in
the north Atlantic for over a hundred years. The Andalusian ship's master
who believed the voyage would normally take three or four months and
that it seemed a miracle for the Armada to complete it in 43 days must
have been a newcomer to the route. A month or five weeks should have
sufficed, and there were Hanseatic sailors in the fleet who would have
known that from direct personal experience. What could not be expected
was that the Armada would now have to endure—in what ought to have
been a favourable season—a voyage that Medina Sidonia was to describe,
with only a little exaggeration, as of 'the greatest travails and miseries
ever seen'.[8]

The account by Fr. Geronimo de la Torre accurately conveys the frustration and privation caused by the unprecedented weather:

We continued our voyage through that sea by Norway, Denmark, Scotland. We sailed as far as the sixty-second parallel, where, in the height of summer, we encountered days so dark, fogs so weird, that our senses were all obliterated. It rained every day, often with downpours that left us sodden. There were chills so cold that they made it seem like Christmas time. No man was willing to go aloft where the pilot was, for all sought a place of shelter. With all this, there was great hunger and great thirst, for our only rations were a pint of water and half a pint of wine and half a pound of putrid biscuit. We were awash with water and it rained down on our heads. Our ship was leaking and we had men dying of thirst. The storms were very violent, for that is a most troubled sea that is always in uproar, and there was no human frame that could endure it. Tempests and heavy seas were so common that scarcely a night went by without wild lurching of the ships. And the nearer we got to home, the worse were the tempests—so much so that many times not I alone but all of us lost all hope that we would ever see land again. We waited only for the ship to be capsized, for there was one time when the foremast was plunged below the water to two yards' depth. Blessed be God Who delivered us from such great adversity.

There was some pardonable exaggeration here. The belief, for instance, that the Armada was blown as far as 62 degrees north seems excessive, though it was common aboard the *San Martín*, fostered, no doubt, by the pessimistic calculations of a pilot. Fray Bernardo de Góngora, for instance, thought that 'we ascended to the sixty-second parallel, where there is no warmth, and I am cold for it was by a miracle that I escaped from Don Pedro de Valdés's ship without clothes, and the duke himself has given me a cloak of his, which is my bed and habit both, to preach the Gospel in'. Fray Gerónimo also suppressed in his account or excised from his memory the few days of calm and of following winds which are recorded in other, more detailed logs. He seems, however, to have been recording his genuine impressions with utter sincerity. It was true, for instance, that the storms and squalls were unprecedentedly heavy, that the rain and cold were unusually intense, and that the Armada, when close to home, was blown back to the far north by an almost unique experience: the tail-end of a tropical hurricane, which has since penetrated to the same latitudes on only one recorded occasion, in 1961. The many contemporary claims that the Armada was beaten, not by men, or not by men alone, but by 'the wind and waves' have often been dismissed as Protestant propaganda, designed as a demonstration of divine hostility to the Span-iards. The facts were exploited by Protestant propagandists for that purpose, but are no less facts for that. They were acknowledged, and variously interpreted, on both sides, but the reality of the weather could not be fabricated, nor its efficacy denied. It defeated the Armada and was, and in the historical record remains, strong enough to dispel many of the alternative explanations of the fleet's failure, as surely as it scattered the ships.[9]

For about the first fortnight of the voyage, the Armada had, perhaps, no more than ordinary bad luck. On 12 August, as the fleet cleared the Forth, the English dropped away. The next day a spell of storms and rough seas began that would last for five days, scattering the English fleet so badly that it gave rise to false rumours of a great Spanish victory at sea. The wind remained in the south, favourable for the Armada's intended course round the north of Scotland. Medina Sidonia, however, was determined to keep his fleet together, and no ship was allowed to run for home alone. In the conditions of iron discipline the duke imposed, with hanging as the penalty for breaking ranks, the ships strained to keep company. On the day the storms began, he ordered the biscuit ration to be shortened to eleven ounces a day and the wine ration to half a pint, with officers, including himself, setting an example. Horses and mules were thrown overboard to save water and sailing instructions were issued which imposed a safe course north of Scotland and out into the North Atlantic, giving the dangerous Irish coast a wide berth and peremptorily forbidding the captains to risk their ships by prematurely seeking havens. These measures, taken together, should have been enough to guarantee the survival of the Armada. First, however, the fleet was sundered by storms, then its course arrested by an unprecedented bout of adverse weather, its discipline eroded by adversity, and its morale dissipated at least in an atmosphere of *sauve qui peut*.

The slow crumbling of the Armada—which later became a rapid cascade—began in the third week of August. It was impossible to keep the fleet together indefinitely for three insuperable reasons: the variable sailing qualities of the ships, which caused the slowest and least weatherly to straggle in increasing numbers from about 17 August onwards; the violent winds that dispersed the fleet whenever Medina Sidonia managed to collect it—or a substantial part of it; and the frequency of head winds, against which many ships could make little progress. There is room for debate on the problems of how much was contributed by indiscipline and decayed morale, and how much by Medina Sidonia's own failure to insist on the maintenance of unity and adherence to his own sailing instructions. Juan de Cardona reported 'that some men say that when those [ships] which are missing left the duke's convoy, it was at his wish. Others say that the duke told everyone to go the way they thought best.' This is borne out, to some extent, by reports from individual ships. The *Zúñiga* was told, for instance, first by the duke and again later by Recalde, that no help could be provided with her broken rudder and that she must get

The most famous of Armada counters. Thunderbolts flash from the name of God. In the foreground ships exchange cannon fire. The legend reads 'Jehovah blew and they were scattered.' On the reverse the True Church is assailed by winds and stormy seas, but the legend proclaims 'I am attacked but unharmed.'

by as best she could. Groups of stragglers whose slowness imperilled the fleet were systematically detached as the voyage proceeded and detailed to bear each other company. But all such compromises with the cohesion of the fleet were forced on the commander by necessity. And, as we shall see, most of the individual decisions to break formation, in the face of the weather of late August and September, and, in many cases, as September in turn began to wane and the weather grew even worse, to set aside the fleet's sailing instructions and seek a temporary haven, were also compelled by extremes of need, rather than being mutinous acts or the craven outcome of collapsed morale.[10]

The first occasion on which the collective security of a single formation seems consciously to have been suspended by the decision of the duke, or of a council of war, was probably on 14 August. The seas were so heavy and visibility so bad that ships could not see each other and needed plenty of sea-room. At this time the weather was already so bad that Fenner expected it to take a heavy toll of the Armada by denying havens, punishing ships, and inflicting unwonted cold on the men. 'Mine opinion,' he told Walsingham, 'is they are by this time so distressed, being so far thrust off, as many of them will never see Spain again; which is the only work of God, to chastise their malicious practices, and make them know that neither the strengths of men, nor their idolatrous gods can prevail, when the mighty God of Israel stretcheth out but his finger against them.' Fenner's language was misleading, perhaps, because of his confusion of natural with providential forces; but his prediction, though premature, was accurate, and an eloquent testimony to the unseasonable severity of the winds and seas. In the circumstances, the commander of the Armada continued to confide in the sailing instructions, but seems to have been willing to compromise on the unity of the fleet. It was a further case of 'obedience to the wind'.[11]

Medina Sidonia allowed the fleet to divide into groups, probably closely based on their squadrons, with the intention that they should reunite when the weather improved. In fact, it only got worse. Three big Mediterranean merchantmen disappeared on that day and never resumed contact with the main fleet: some of their company must have got to Norway, from where 300 fugitives were later shipped home via Scotland; at least two of the ships must have got as far as Ireland, because no more than one ship of the Levantine squadron is undocumented beyond this point. On the 17th, during the night, the Armada encountered the worst storm yet and Pedro Coco's ship, with some other hulks, lost contact with the main fleet until the 19th. The severity of this blow was confirmed by Drake: 'Her Majesty's good ships felt much of that storm and lost many of their boats and pinnaces.' On the 19th, though somewhat improved, the sea was still too heavy for Pedro Coco to distribute sick rations to the *San Marcos*, though he did manage to get some delicacies aboard the *San Juan* of Recalde. Among ships missing from the reunited fleet were two of the Mediterranean-built fighters that had distinguished themselves on 8 August—the *Rata Encoronada* of Don Alonso de Leiva and the Ragusan *San Juan de Sicilia*, which had barely survived the battle. Another tempest that night dispersed the fleet again. Pedro Coco's hulk tried to stay with Recalde, but, forced, as usual, by her bulky build, to run with the wind, she found herself alone and unable to recover the fleet for two days.[12]

Meanwhile, however, the worst period seemed to have passed. The fleet had been able to steal through the sea between the Faroes—or perhaps even the Orkneys—and Shetland, whereas Medina Sidonia had expected to have to steer right round the outermost of the northern islands. This deft piece of navigation had been negotiated with a following wind and without loss or mishap to any ship. On 21 August, Medina Sidonia was able to report himself off the north-east coast of Scotland with a favourable wind for Spain. There seems to have been a general optimism aboard the fleet that with a little more rationing to see them through the belt of westerlies, their troubles would be over and Spain soon in sight. Don Francisco de Bobadilla—never one for complacency—succumbed to the hope of a happy ending: at this stage, he could even suppose that the fleet might be sent off at once again to Ireland after re-victualling. 'The soldiers will be fit to serve if they give them food and do not let them suffer from hunger.' Fray Bernardo de Góngora managed a little self-congratulation. Though 'the outcome has not been what we wished', the duke 'will have done no mean thing in restoring the Armada intact to the king after dealing harshly with a hundred of the ships of our enemies, who, to be sure, will not be left speaking well of us'. These men and others aboard the flagship and around the fleet busily wrote hurried dispatches and letters home to be carried by Don Baltasar de Zúñiga aboard the fast pinnace the duke prepared to send ahead to Spain.[13]

Their hopes were about to be cruelly disappointed. The wind veered again and by the 24th the Armada was still, by general consent of the experts on board, struggling at more than 58 degrees of latitude north. What was worse, the fleet was disintegrating. Medina Sidonia's careful custodianship of his ships was breaking down in the face of furious head winds that blew them apart at uncontrollable rates. The two main groups—one of about fifteen ships with Recalde, the rest with the flagship—were still in touch, but individual ships had lost contact and were feared or hoped to have cut and run for havens in Iceland and the Faroes.

At this time, to judge by Pedro Coco's account, the question arose anew of whether the entire fleet should not seek a harbour in Scotland or Ireland. The apparent impossibility of making headway, while rations grew short, must have suggested that the duke's sailing instructions would have to be abandoned. Diego Flores was said to favour this desperate expedient, but the bulk of expert opinion adhered to the view that these coasts were too dangerous and the westerlies too persistent. The duke confirmed that they would 'avoid Ireland by all means'; more stinted rice rations—fifty pounds per ship—were doled out to the sick and wounded, and the fleet made to struggle on.

'From 24th August to 4th September,' Pedro Coco's ship 'wandered lost amid storms, fogs and downpours, and as this hulk cannot tack against the wind, and we had to go where the sea took us, we could not see the rest of the Armada until the last day when we rejoined it.' For the next six days of heavy seas and fog, they managed to keep generally within sight of other ships by constant hauling on the bowline. These early days of September brought some of the least inclement weather apart from a severe gale reported by some ships on the second of the month. On 3 September, Medina Sidonia wrote to the king in a mood of nicely mingled hope and foreboding, with ninety-five ships still clinging to the main fleet.

The wind has lengthened, more favourably, for the west-north-west. But the winds in these parts are so violent and stormy, and tend to prevail from the south, so that there can be little certainty that the present wind will last. May Our Lord in His goodness give us fair weather, so that this Armada can reach a port soon, for we are running so short of victuals that if this voyage is much prolonged, for our sins, everything will be lost irremediably.

The *Gran Grifón*—now lagging badly behind, tenuously in contact with some other stragglers, including the *Trinidad Valencera*—reported good weather from 5 to 11 September, before a storm ahead drove her right back to an estimated 54 degrees north. Alonso de Vanegas, whose 'log', apparently written later and partly from memory, cannot be trusted on precise dates, also recorded most of the first week of September, until the 6th, as favourable to the progress of the fleet. Indeed, at about this time, the English were beset by fears that the Armada might be about to return to challenge them in fight again. The report was brought by Sir Edward Norreys the very day after Drake and Hawkins had discharged a large number of ships. Burghley was incredulous but the commanders took no chances and now stirred painfully to rearm to face the danger. Drake expected the Armada to 'jump with fair weather'. Seymour, disenchanted with the tedium of the service and the lost opportunities of the war, recovered a little of his old animation. 'So long as there is an expectation of the Spaniards to return, I would not have the thought once to return before some better service be accomplished.'[14]

English fears, of course, were as illusory as Spanish hopes. With the exception of a few stragglers who had already given up the struggle and gone off in search of refuge, the Armada was bound purposefully, and still with a surprising degree of cohesion, for Spain. Now, however, with that goal almost within reach, the fleet was struck, repeatedly and progressively, by the worst weather yet encountered. The first bout of this seems to have occurred at about the end of the first week of September, when the English in the Channel felt 'a wonderful storm' and 'extreme' and 'frightful' weather. This could well correspond to the period of two days endured by Alonso de Vanegas aboard the *San Martín*, from 6 to 8 September, when 'we had adverse winds and a tempestuous sea with visibility so bad that we could hardly see one ship from the next'. There seems to have been another terrible blow on about the 12 and 13 September, when the flagship was thrust back, so far north that, according to Vanegas, the cold took a heavy toll of the Negroes and mulattos aboard. In this period the ships keeping company with the main fleet fell abruptly from the ninety-five recorded on 3 September to no more than sixty by about the 17th, not only because of the natural dispersal caused by the severity of the storms but also because the continual frustration of the fleet's efforts to make headway was eroding both rations and morale. Most of the ships that disappeared broke ranks deliberately or despairingly, seeking Irish havens in defiance of their sailing instructions or driven westward by storms and gales. The ships that maintained discipline—a little less than half the original strength of the Armada—were rewarded with the weather's most capricious stroke of cruelty yet, when they were struck, over a period of four days from 17 September, by a veritable hurricane. On the first of only two recorded occasions in the last half-millenium, the tail of a tropical typhoon lashed out across the Atlantic; it caught the Armada when the fleet was at an estimated 45 degrees north, sensing the proximity of Spain. Even the *Gran Grifón*, straggling far to the north, was almost overwhelmed by it. For Medina Sidonia it was 'so great a tempest that we all thought to die'. For Vanegas, more graphically, 'so great a tempest and chance stroke of wind and sea that it seemed in that hour that it would overwhelm and destroy the whole world'. The remaining unity of the Armada was shattered. When the wind left it, there were only eleven ships together. When the fleet struggled into harbour four days later, the master of a Sevillan bark counted sixteen sail.

A group of twenty-one reached Laredo. Medina Sidonia was told of five or six that were in Biscayan ports; nine or ten more joined him at Santander, or were thought to be approaching, during the 23rd. Unknown, at first, to the duke, ten others arrived at San Sebastián. The effect of the hurricane was therefore, it seems, to scatter, rather than to sink, the ships. For many of those, however, which had already been left straggling, far from Spain, the effect was fatal.

Others struggled in during the following weeks—Juan Martínez de Recalde in the *San Juan* reached Corunna on 7 October; Martín de Bertendona in the *Regazona*—one of only two great ships of the Levant squadron to return home—followed on the 8th, floundering into El Ferrol without a rudder. The arrival of Juan Gutiérrez Garibay in *Nuestra Señora de Begoña* was not reported till the 14th. Within a day of the first arrivals, as many ships as had been with Medina Sidonia when the last bad weather struck all seem to have made port. Over the next three weeks, other vessels arrived which had long been separated from the flagship. For many ships that had been straggling well to the rear, however, pounded by storms for most of the month of September, the effect of bad weather was cumulative, and fatal, as we shall see in the next chapter.[15]

Of those ships that did return, moreover, the condition reflected the severity of the experience to which they had been exposed. The flagship had lost 180 men through sickness, including three of the four pilots aboard, and of Medina Sidonia's sixty personal fighting retainers only two, he claimed, were still fit. The *San Martín* arrived 'so rent by great shot that the pumps have to be worked day and night on both sides, ceaselessly'. Medina Sidonia warned Juan de Idiáquez, 'Your Lordship will not be able to believe the state in which the ships are arriving'. The state of the Duke of Tuscany's galleon, summarized by Francesco Bartoli, the captain's nephew, in a petition to the king, is probably exaggerated—as it is in effect a claim for compensation—but is typical of the terms in which returned vessels were described. 'She is to be found', he wrote,

in the port of Santander, all pierced by the enemy's artillery, and, because of the tempests to which she has been exposed, without boats, sails, anchors or the instruments she needs, and her crew wasted without having anything to eat, so that the ship is not just unable to sail, but in extreme danger of foundering in the very port.

The Guipúzcoan ships arrived 'broken and distressed, without rigging, sails or cables'. The *Doncella*, which some experts believed to be sinking on 9 August, finally did so in harbour two months later.[16]

More pitiful even than the plight of the ships was that of many of the

The passing of the Armada inspired English pride and joy, sumptuously evoked by this gold medallion. On the obverse the queen is shown in unsurpassed magnificence. The rigid posture, the upheld orb and sceptre, and the rich setting all convey an impression of majesty. Elizabeth is surrounded by the punning legend 'No other circle is richer in all the world.' On the reverse the circle of the sea protects a stylized island just recognizable as England—much squashed and foreshortened—from the contours of the coast. Cities and shipping flourish in peace; the seas teem; and the sun emerges from the receding clouds of war. The green bay tree rises, unruffled and erect, over the legend 'No very peril harms me.'

A few surviving artefacts, such as this cast-iron rustic fireback, seem to attest the depth of the contemporary popular impact of the thwarting of the Armada. In conjunction with the date 1588, the prominent anchors with their coiled cables in the midst of a flourishing garden appear to convey, albeit less sumptuously, the same message as the gold medallion.

men at every level of command. They returned weary, as Bernabé de Alvia reported, 'from the great travails they have suffered, for they have endured much hunger and thirst. . . . After the long voyage they are naked. The sailors have suffered more from the labours they have had to perform than the soldiers.' The last point tended to be confirmed by open signs of enmity between the two services, the army men claiming they had manned the ships, the seamen protesting that they had been slighted and maltreated by the soldiers. According to Oquendo, the men in San Sebastián in October were 'dying like bugs', pitiful to behold, 'naked and without so much as a shirt'. The hospitals, which, as we shall see, were quickly mobilized in the receiving ports, were full. At the most conservative estimate, there were a thousand sick in Santander alone, and probably well over three thousand in the fleet as a whole.[17]

Though the English ships, as we have seen, had been largely spared from serious damage, this sort of personal suffering embraced the English forces with almost equal ferocity. This can be observed, first, in the ranks of the high command. On the Spanish side, while Oquendo, Recalde, and Bartoli were among the captains who survived the voyage only to expire, exhausted by hardship, on their return, on the English side the same fate befell the Earl of Leicester, who had worked and worried indefatigably in command of Elizabeth's land forces and been 'huntsman, caterer and cook' in his own camp to boot. On 11 August Drake predicted that Medina Sidonia would soon 'wish himself at St Mary Port among his olive trees'. With allowance made for the fact that the duke's seat was at San Lúcar de Barrameda, not Puerto de Santa María, the prediction was to be exactly fulfilled. The duke received his *congé* on 29 September on the fully justifiable grounds of health broken by the experience of the Armada: 'since you say', as the king wrote, 'that it is important for your con-valescence not to stay in those ports nor spend the winter in a cold part of the country, but go now to be cured and recover your strength in your own land'. Medina Sidonia's own plea to be released had been couched in somewhat stronger terms:

And as for affairs of the sea, in no instance and in no way will I ever have anything to do with them again, even if it cost me my head, which would be a better way to go than to die in a post of which I have no knowledge or grasp and in which I am obliged to have faith in those who advise me without knowing what their motives are. I am so weak that my hand cannot write these words, nor can I add any more.

Disillusioned as Medina Sidonia was, however, his aversion from the business in hand was hardly greater than that voiced on the English side by Lord Henry Seymour, who begged Walsingham, 'Spare me not while I am abroad. For when God shall return me, I will be kin to the bear, I will be haled to the stake, before I come abroad again.'[18]

Seymour had particular reasons for his dissatisfaction: he felt, as we have seen, cheated of a share in some of the most potentially glorious passages of the campaign; he had been frustrated at what he saw as the waste of his time and service guarding the Narrow Seas against an attack he was convinced would never come. Yet it remains a curious fact that disillusionment, recrimination, and mutual reproach were almost as rife on the English side after the Armada, amid celebrations of success, as on the Spanish side amid a consciousness of failure. In the late autumn and

The Earl of Leicester as Governor-general of the Netherlands. Though attired for peace, he has his arms about him and a gorget at his neck, while a military camp is displayed beyond the window. As in Burghley's portrait on p. 73, considerable prominence is given to the Order of the Garter, which the earl wears below his knee while pointing to the insignia which rest on the table. Within the Garter an image of St George appears—a considerable embarrassment to Protestant iconoclasts and opponents of saints' cults. According to John Rainolds in 1596 the image served to incite the Garter knights to fight the dread dragon, the Antichrist of Rome.

winter of 1588 the Spaniards had an opportunity to mobilize their superior resources of efficiency and organization to cope with the human and material debris of their fleet. The English, meanwhile, proved incapable of exploiting their advantage or of caring for the men they had in arms.

The Armada had been expected to return to Corunna and there supplies had been stockpiled and accommodation prepared to receive the home-coming fleet. The Archbishop of Santiago had bought meat, clothing, and medicines for the sick and had ordered the physicians and surgeons of his hospital in Corunna to be ready to tend them. In the event, Spain was to be doubly surprised—first, at the scale of need created by the arrival of the fleet, particularly for the care of the large numbers of sick and wounded; secondly, by the Armada's dispersal around northern ports. As Medina Sidonia explained to the king, he had been heading for Corunna, but the violent storms he encountered in the last days of his voyage disposed otherwise. The king and the realm, the bishops and bureaucrats who were made chiefly responsible for the reception of the returned Armada, responded with astonishing speed to the new situation. Don Baltasar de Zúñiga arrived with the duke's messages only three days before the first contingents of the fleet, yet within a day of Medina Sidonia's arrival the first relief column was threading its way by mule train across the mountains from Burgos. Lights were erected on the northern coast to guide remaining stragglers. The provisions collected in Corunna were redistributed to other ports, where they would be most useful, and pleas went out from the king to every potential source of help in the kingdom, in order of proximity to the remnants of the fleet. Burgos was first in the field with aid for the stricken Armada because of her long-perfected communications with Santander and other northern ports, which exported her wool, and because she had been designated in advance, together with Santiago de Compostela, as a centre for the organization of supplies to the Armada. But the other major cities which were best placed to serve, Logroño and Valladolid, were not laggard. It took the Council of Logroño only nine days to commission and receive a report on the state of the fleet and the nature of the most urgent needs and to dispatch a thirty-strong mule train with 1,533 lb. of basic rations, 520 lb. of almonds, 525 lb. of sugar, 289 lb. of the preserves that were thought to be good for invalids, 2,501 lb. of eating apples, 479 lb. of cooking pears, 619 lb. of pomegranates, 1,146 pints of sweet oil, 1,305 bolts of linen for bandages, 150 pairs of leather shoes, 72 pairs of hose, 100 baskets of bootlaces, 30 counterpanes emblazoned with the arms of the city of Logroño for the hospital beds, and 500 ducats in cash with which to buy mattresses. Valladolid was quick to respond to the king's demand, received on 4

October, for immediate aid to the fleet. A wagon train painted in the city's livery was loaded with preserves, pumpkins, sweet biscuits, quince jelly, good white wine, medicines, and 'luxuries' and 'doctors and grave and learned surgeons of this city, each one duly examined to establish that they are persons practised in their art'.[19]

Over the next few weeks, pledges to aid the king in his necessity flowed in from all over the realm. Towns contributed according to their capacity. Madrid and Valladolid, rivals for the privilege of housing the royal court, offered 100,000 ducats each; Segovia promised five hundred infantry, Huete 250 arquebusiers and 12,000 ducats. Magnates and prominent churchmen offered their persons and fortunes, often without limitation of term 'for the duration of the war'. Almost all such pledges referred explicitly to the circumstances of the return of the Armada—'seeing the present state of things' or 'having heard the news of how His Majesty's Armada has arrived in the ports of this kingdom' or 'the outcome of the Armada being known', or, in the case of Ávila's pledge, 'having perceived the Catholic and holy zeal with which His Majesty has begun and wishes to prosecute the campaign against England'. At the Cortes celebrated in Madrid early in the new year, the deputies recognized the need for heavier taxation to cope with the aftermath of the Armada, 'for therein rests no less than the safety of the sea and of the Indies and of the Indies fleets and even of our own homes'.[20]

Elizabeth might almost have envied Philip his failure. The plea of necessity, which was so effective a regulator of a monarch's relations with subjects, and so powerful an instrument for the fiscal exploitation of a realm, was not available in a country which had just had a salutary escape and claimed a great victory. Yet England's immediate real needs were very similar to those of Spain. There were huge forces to be fed, paid, and demobilized and large numbers of sick to be tended. Howard had been apprehensive about the health of his men since before the start of the campaign. 'God of His mercy keep us from sickness,' he exclaimed to Walsingham on 23 July, 'for we fear that more than any hurt that the Spaniards will do.' These fears were soon realized. On 20 August, he reported to Burghley, 'Sickness and mortality begins wonderfully to grow amongst us; and it is a most pitiful sight to see, here at Margate, how the men, having no place to receive them into here, die in the streets'. Here was an irony worthy of Cuellar's pen: the victorious English dying in the gutter; the defeated Spaniards going home to hospital beds and embroidered counterpanes. Howard claimed that the *Elizabeth Jonas* had lost 200 men out of a company of 500 within three weeks or a month in Plymouth: if so, pestilence was indeed a worse foe than Spain, though both figures are too high to be readily credited. He had taken out her ballast and made fires of wet broom for three or four days, but the infection lingered, perhaps, Howard thought, 'in the pitch', while her sick were consoled by a dole of £7, divided between them, from the Pipe Office. Nor was sickness the only problem. The Lord Admiral begged for pay, reinforcements, and above all fresh clothes—hose, doublets, shirts, and shoes—'for else in very short time I look to see most of the mariners go naked'.[21]

The English fleet and, to a lesser extent, the army, were suffering from much the same combination of adversities as faced the Armada. They too—or, at least, a considerable proportion of them—were obliged to remain in arms and on their ships long beyond the range of their rations. They, too, were battle-weary and, after the August storms, weather-beaten and widely scattered, though of course the intensity of their sufferings was modest by comparison with what the Armada had endured in the fight and would undergo in the Atlantic. By the end of August, Seymour was reporting his men 'dropping away' by reason of 'the cold nights and cold mornings we find' and by the beginning of September the sickness had become ubiquitous in the fleet, at a time when, as we have seen, apprehensions were rife that the Spaniards might be about to return. It was therefore with some urgency that Howard wrote to the Council.

As I left some of the ships at my coming up, so I do find, by their reports that have looked deeply into it, that the most part of the fleet is grievously infected, and die daily, falling sick in the ships by numbers; and that the ships of themselves be so infectious, and so corrupted, as it is thought to be a very plague; and we find that the fresh men that we draw into our ships are infected one day and die the next, so as many of the ships have hardly enough men to weigh their anchors; for my Lord Thomas Howard, my Lord Sheffield, and some five or six other ships, being at Margate, and the wind ill for that road, are so weakly manned by the reason of this sickness, and mortality, as they were not able to weigh their anchors to come as we are.

Despite the current danger, Howard saw no help but to lay up half the ships and put the men ashore. 'My Lords,' he continued in justification of his decision, 'we do not see, amongst us all'—referring to his council of war—'by what other means to continue this service; for the loss of mariners will be so great as neither the realm shall be able to help it and it will be greater offence to us than the enemy was able to lay upon us; and will be in very short time answerable to their loss, besides the unfurnishing of the realm of such needful and most necessary men in a commonwealth.'[22]

Howard turned to the decaying morale of his men and the flagrant

Most gratious Soueraine Lady, The God of heauen and earth,
(who hath mightilie, and euidently, giuen vnto your most excellent
Royall Maiestie, this wunderfull Triumphant victorie, against
your mortall enemies) be allwaies, thanked, prayfed, and glorified:
And the same God Almightie, euermore direct and defend your
most Royall Highnes from all ebill and encumbrance: and finish
and confirme in your most excellent Maiestie Royall, the blefsings,
long since, both decreed and offred: yea, euen into your most
gratious Royall bofom, and Lap. Happy are they, that can
perceyue, and so obey the pleafant call, of the mightie Ladie,
OPPORTVNITIE. And, Therfore, finding our duetie concurrent
with a most secret beck, of the said Gratious Princefs, Ladie
OPPORTVNITIE, NOW to embrace, and enioye your
most excellent Royall Maiesties high fauor, and gratious great
Clemencie, of CALLING me, Mr Kelley, and our families,
boame, into your Brytifh Earthly Paradise, and Monarchie
incomparable: (and, that, abowt an yere since: by Master
Customer Yong, his letters,) I, and myne, (by God his fauor
and help, and after the most conuenient manner, we can,)
Will, from hencefurth, endeuour our felues, faithfully, loyally,
carefully, warily, and diligently, to ryd and vntangle our
felues from hence: And, so, very deuowtely, and Sowndlie,
at your Sacred Maiesties feet, to offer our felues, and all,
wherein, we are, or may be bable, to serue God, and your most
Excellent Royall Maiestie. The Lord of Hoasts, be our
help, and Gwyde, therein: and graunt vnto your most excellent
Royall Maiestie, the Incomparablest Triumphant Raigne, and Monarchie,
that euer was, since Mans creation. Amen.

Trebon. in the kingdome of Boemia,
the 10th of Nouebre: A. Dni: 1588: style veter.

Your Sacred and most excellent
Royall Maiesties
most humble and dutifull
Subiect, and Seruant:
John D

The magus John Dee, then seeking the patronage of the esoteric Emperor Rudolph II in Prague, was in no doubt when he heard of it in November 1588 that the outcome of the Armada represented a triumph for England. The news inspired him to think about returning home 'into your British Earthly Paradise and Monarchy incomparable'. It must be admitted, however, that Dee had an ulterior motive. His bid for imperial patronage had ended in a decree of expulsion as a fautor of heretics and though the sentence was only partially enforced Dee's continued presence in Bohemia was precarious.

want of pay in terms that foreshadowed the cries of anguish of the returned Spaniards. Oquendo, for instance, was to refuse Medina Sidonia's order to transfer the contents of his pay ship to Santander on arrival in Spain because 'it is surely a shame to behold the men of these ships, so sick, so naked and without a shirt, and the hospitals full, and if they see this money transferred they must all expire'. Towards the end of October, the accountant García de Cerralbo warned the king, 'They are suffering such extreme need and are so disgusted at not being paid that I believe it is impossible to go on sustaining them'. To judge from Howard's account to the Council, the English had already got to that stage at the beginning of September: 'My Lords, I must deliver unto your Lordships the great discontentments of men here, which I and the rest do perceive to be amongst them, who well hoped, after this so good service, to have received the whole pay, and finding it to come but this scantily unto them, it breeds a marvellous alteration amongst them.' Until the ships were finally discharged, in mid-September, Howard continued his efforts to 'open the Queen's Majesty's purse something to salve them' ... on the grounds, expressed in a letter to Walsingham on the 8th, that 'it were too pitiful to have men starve after such a service. I know Her Majesty would not, for any good ... for we are to look to have more of these services; and if men should not be cared for better than to let them starve and die miserably, we should very hardly get men to serve.' Demobilized English soldiers from the camp at Tilbury were said to be selling their armour with complaints that they had got no pay. There were some curious anomalies. Edward Norton, gentleman, who is commemorated on a tablet in St Bartholomew's Church, Hyde, Hampshire, received a pension of 2d. per day for his 'good service by sea'. There were extraordinary doles on the English side for the wounded, who got £80 between them, and some of the sick and the 100 men who manned the fireships—£5 collectively for them. On the Spanish side, the Basque seamen complained because their Andalusian colleagues got their pay first (though the complaint may have been baseless). In any case, they were disgruntled at the level of 'the very small pay they are given, which is the old one which was fixed fifty years back, whereas at the present time there has been an increase of a quarter, at least, in all thing needful for human life and in materials for shipbuilding'. Pay was the most sensitive issue for the majority of men on both sides who were still reasonably fit after the rigours of the campaign, and for their communities at home who were impatient for the economic benefits of their pay; the port towns of England, no less than those of Spain, spent the winter pressing their government to settle accounts.[23]

Despite the common experience of England and Spain in the aftermath of the Armada, the Spaniards confronted one problem which the English did not share. No English ship had been lost; even the worst damage was superficial. But the Armada, scattered, distressed, and debilitated, seemed—so the man appointed by the king to take it in hand thought—to have 'ceased to exist'. At least, he wrote, referring to the prophecy that the 'year of the eights', presumed to be 1588, would be marked by great achievements, 'I shall look on the Armada as in abeyance until the year 1590. I am bound to think that the year of the eights, so ardently expected, will turn out to be 1800.' Pessimism was tolerable, perhaps, if garbed in this sort of grim humour. Generally, however, it had to be resisted, if morale was to be salvaged from the wreck of the fleet. The Armada was facing the dilemma of Don Quixote, whose good deeds were all frustrated and whose good intentions turned to ashes. It requires special pleading, in such circumstances, to summon up the spirit to persevere, or to perceive redeeming features in what looked like an unmitigated disaster.[24]

Medina Sidonia, however, took a different view, shared by many of the survivors and, to some extent, by the king himself:

We can say that it has been a special miracle of God's to have led the Armada to these havens, the fleet burdened with miseries and disasters, the ships full of shot and ravaged by the fury of the storms, without a shred of rigging, without sails or masts, letting much water, and without crews—since most of them are dead or sickly. And were it not for the soldiers who remain healthy there would be no one to man the sails.

In other words, though the afflictions of the Armada were remarkable, its endurance was more remarkable still—indeed, miraculous. There was much to be said for this point of view. In its fighting and sailing record alike, the Armada had performed well. If its losses are considered in detail, in the context of the exceptional conditions experienced, they seem surprisingly modest overall, and to have occurred surprisingly late in the day.

In reviewing the Armada's losses, we should leave some categories of shipping aside. Vessels which dropped out of the fleet before it reached the Channel cannot usefully be taken into account. The hulk *David*, for instance, and perhaps a few other ships, were left in northern Spanish ports after the June storms showed their fragility. The four galleys failed to complete the crossing of the Bay of Biscay, and one was wrecked at Bayonne. The *Santa Ana*, which was to have been the flagship of Recalde's squadron, developed sailing difficulties and withdrew from the fleet, putting into Le Havre for repairs. There she spent months in dock, twice fighting off English attempts to break into the harbour and sink her. All these vessels, if added to the figures for Spanish losses, would inflate them unfairly. Similarly, the other ships that were to end their journeys in French ports—the galleass *Zúñiga* and the hulk *San Pedro Menor*—are marginal cases, which can be counted as having escaped. And it is prudent, when computing the losses, to limit the enquiry to reasonably well-documented ships. In practice, this means the big ships—the fighting vessels and transports. Pinnaces and tenders are, in most cases, not individually documented: while there are good reasons for believing that up to five may have been wrecked in Ireland and that others may have foundered at sea, the specific fate of so few is recorded, and the records are so obviously incomplete—the Ragusan-built *San Jerónimo* (*Sveti Jerolim*), for instance, is recorded as missing but can be assumed to have got home from the evidence of the survival of individuals on board—that it is prudent to leave the problem of accounting for them on one side. This makes little difference to an overall assessment of the fate of the Armada, since the costs in terms of lives, equipment, and resources were relatively small when pinnaces were lost, with ship's complements usually of less than fifty, compared with the big ships in which huge sums were invested, aboard which hecatombs of sailors and warriors died and fortunes in stores and equipment were expended. The two large pinnaces, the *Julia*

The construction of a pinnace, from Diego García del Palacio's manual of Atlantic navigation and naval warfare published in the year before the Armada.

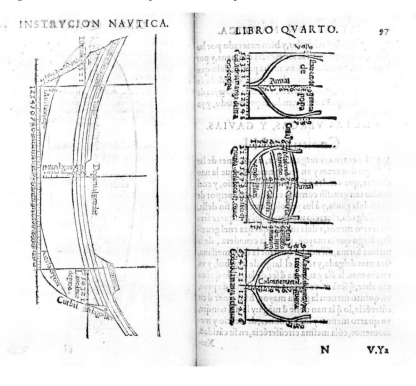

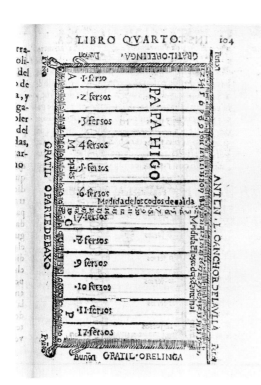

Diego García de Palacio's design for the sail of a pinnace. The small roman numerals give the dimensions in cubits. Notice the loops and bolts for the rigging.

and *Augusta* of the squadron of Portugal, with ship's companies of 135 and 92 respectively, both got home. Of thirty-three smaller pinnaces named in the Order of Battle, twenty appear listed as missing in the musters drawn up in October 1588, but this is no ground for assuming that anything like that number was genuinely lost.

It is instructive to classify the losses in three ways: by squadrons, or type of shipping, by the means of loss, and by the date of loss. The Andalusian squadron proved outstandingly durable. Of its ten great ships, eight reached home safely. Of the two vessels lost, the flagship, *Nuestra Señora del Rosario*, was disabled by a collision in the Channel and captured by Drake: this is not necessarily any reflection on the fighting or sailing qualities of the vessel, which were never put to the test, though her English captors complained of her, claiming that she needed constant pumping to keep her afloat in dock. The other lost Andalusian was the *Duquesa Santa Ana*. She was apparently regarded as the least weatherly vessel of her squadron as she was the only one to be classified as a hulk. On the other hand the nickname *Duquesa* was said to be a compliment to the commander-in-chief's wife. Unless the compliment was backhanded, she must be supposed to have been an estimable vessel and was certainly a large one, with 23 guns and a company of over 50 men. There is no

evidence that she was seriously battle-damaged: her only enemy, therefore, was the weather, which she withstood impressively—if allowance is
made for her classification as a hulk, which implies poor handling qualities
in head winds. Not until 17 September—probably the worst day, that is,
of the tropical storms—did she seek refuge in Ireland. Not until the 29th
was she wrecked, and then only as a result of grounding through the loss
of her last usable anchor.

The other four squadrons of Atlantic-built fighting ships and converted
merchantmen fared only slightly worse. The Guipúzcoan, Biscayan, Portuguese, and Castilian squadrons lost three great ships each out of their
effective strength during the voyage—thirty per cent of the strength of
the Guipúzcoans and Portuguese, rather less in the Castilian case, since
the squadron had thirteen great ships altogether, with the converted
merchantmen and the *San Juan* of Fernando de Hora. Of the Guipúzcoan
losses, one can be ascribed to accident or sabotage—the explosion and
firing of the *San Salvador* in the Channel. Of the others, the *Santa María
de la Rosa*, was a notoriously weak ship, which had already been dismasted
once, in the June storms, before the Armada finally set sail. Her new mast
lasted well enough to go down with the rest of the ship on a reef in the
Blaskets on 16 September 1588 but it was, in any case, poorly designed
and poorly positioned, too far for'ard for the overall balance of the ship.
As with the *Duquesa*, weather-beating alone, without significant help from
battle damage, seems to have been the undoing of the *Rosa*. The last
Guipúzcoan wreck was probably the *San Esteban*, which, by a process of
elimination, can be identified as one of two ships lost in Ireland on 20
September, but nothing is known of how or why she got into difficulties:
the date of the wreck suggests she was another victim of the storms from
the tropics. The *Doncella*, which survived all the Atlantic weather, to
founder, after reaching home, in the harbour of Santander, was a Guipúzcoan classed as a hulk: but she deserves to be ranked with the survivors;
in view of her unweatherly build, and the fact that she was battle-damaged
too (as can be inferred from the rumour that she was already in a
sinking condition on 9 August), her endurance was too remarkable an
achievement to go unrecognized.

The Portuguese squadron was distinguished by having great ships
exclusively of the galleon class, all Portuguese-built but for the Duke of
Tuscany's galleon, which was constructed along 'Atlantic' lines. It also
included four of the half-dozen probably most battle-damaged ships in the
fleet after the fight of 8 August. Of these four, only the *San Martín* returned
home. Even with the one-way pounding at short range described in the
previous chapter, the *San Felipe* and *San Mateo* remained afloat until they

The explosion aboard the *San Salvador* is dramatically rendered after the House of Lords tapestries by John Pine. The force of the blast strews the sea with jetsam and tosses bodies into the air. All Pine's engravings seem to combine romantic flourishes of that sort with on the one hand meticulous realism in the drawing of the ships, and on the other the curious intrusion of traditional features like the prolific whales and sea monsters and the conventional waves in the background.

drifted on to the Nieuport banks, but, as the English gunnery effectively disabled them, they must be counted as battle casualties. The *San Marcos*, probably abandoned or wrecked in Ireland, perhaps on or about 20 September, was a long-term victim of enemy action: to judge from the performance of comparable ships, she was unsinkable by English gunnery alone and would, if in good condition, have been invulnerable even to the terrible September storms. Severe damage from both sources together, however, was enough to make an end of her. It is impossible, other than by an appeal to chance, to explain why the *San Martín*, which seems to have taken an even harder pounding in the firefight of 8 August, fared better than her sister ship. It is worth pointing out, however, that Medina Sidonia kept the *San Martín* scrupulously to his sailing instructions, which stressed the overwhelming importance of avoiding the Irish coast. If the *San Marcos* was wrecked, rather than abandoned, she was a victim of the fulfilment of Pedro Coco Calderón's doom-fraught oracles.

The only other galleon to be lost was the *San Juan Menor* of the Castilian squadron, which formed part of a small flotilla that turned aside for Ireland on 20 September and was dashed to pieces on Streedagh Strand on the 25th. It is a mystery why this galleon—and the other *San Juan*, vice-flagship of the same squadron, which eventually reached home safely, should have been among the Armada's stragglers. Apart from the *San Marcos*, no other galleons reported serious sailing difficulties, battle damage notwithstanding, or found any need to try to shelter from bad weather on the way. The squadron's other two lost ships, the *Trinidad* and the *San Juan Bautista* of Fernando de Hora, though not galleons, were useful fighting vessels, each, like the *San Juan Menor*, of 24 guns. The *Bautista* did not perform badly: she remained on course until 20 or 21 September and it seems only to have been storm-damaged rigging and the loss of a mast that forced her to try to find refuge in Ireland, where she came to grief on a reef in the Blaskets. The *Trinidad* is too poorly documented to justify further comment, save that she may have been the vessel of that name that accompanied the vice-flagship of the squadron, which had lost her foremast, between 26 August and 12 September. In general, it seems fair to suggest that, despite selective weaknesses, the Castilian squadron was well adapted for the battles against the English and the weather alike.

The last of the Atlantic-built squadrons, that of Biscay, certainly lost two great ships, and almost certainly a third, out of ten. The *María Juan* was the only ship in the fleet to have been reduced to a sinking condition by English gunnery (though even her demise was assisted by the late squall of 8 August). The second lost ship was probably the *San Juan*: her homecoming is not specifically recorded, though the fleet contained so many homonymous ships that the possibilities of confusion are legion. She may have been one of the wrecks, reliably reported but not easily identified, in County Clare between 16 and 24 September. The vice-flagship of the squadron, the *Gran Grín*, of 28 guns, went aground on Clare Island, probably towards 24 September. All three vessels, therefore, seem likely to have been victims of the September storms.

If the Atlantic-built fighting vessels are looked at collectively, and the *Doncella* and *Santa Ana* omitted, eleven great ships can be counted as losses while at last thirty-seven survived. Three were lost directly or chiefly to enemy action, two to accidents, one to a combination of English gunnery and bad weather, and the remaining five directly or primarily to the weather. Of the weather's victims, all seem to have survived well into the period of the worst weather. And only one, at most the doubtful Biscayan—can be suspected of having foundered at sea, the rest being clearly identifiable—as we shall see in the next chapter—with known coastal wrecks. This suggests that important ancillary factors in their undoing were the lack of a port of refuge for the Armada in northern waters, which induced ships to take unacceptable risks on the Irish coast, and the imprudence of departing from Medina Sidonia's sailing instructions: among Atlantic-built great ships, it was only or almost only (allowing for the possibility of a vessel lost in the open sea), those which laid aside their orders and looked to Ireland for help which failed to get home in the end.

The record of ships of Mediterranean or Baltic build and of hulks of uncertain provenance was very different. Eight of the ten great Levanters, two of the four galleasses, and up to eight or nine of the twenty-two ships which probably sailed with the squadron of hulks, disappeared in the course of the voyage. Even the performance of these classes of shipping was in some ways surprisingly good. Of the galleasses, all played a prominent part in the fighting without suffering serious damage. The *San Lorenzo* was the victim of an accident which dislodged her rudder in Calais roads; the *Girona* was the longest-surviving of the ships known to have been lost, and it was again the galleasses' general curse, a weak rudder, that seems to have put paid to her. Since it had been universally believed before the voyage of the Armada that galleasses would be seriously at risk in northern seas, and would need to have access to shelter in French ports in the event of bad weather, it seems astonishing that they should have endured so much and that half the squadron should have remained afloat, albeit in dire plight, at the end of the campaign. The big Levanters were also reputedly fragile, but were evidently more so to storms than to enemy action. The *San Juan de Sicilia* and *Trinidad Valencera* were prominent in the fight of 8 August and the former was one of the most badly battle-scarred ships; the *Rata Encoronada, Anunciada,* and, implicitly, the *Lavia* all sustained some battle damage. Yet, though all those ships became stragglers in the course of the Armada's journey home, it seems to have been their unweatherly qualities rather than the effects of battle that caused them to be detained. The *Regazona,* despite a gallant battle record, struggled home late—more than a fortnight after the flagship—but safe. The *San Juan de Sicilia* succumbed only to sabotage. The *Lavia* survived until 25 September and was one of the last ships to abandon the struggle to get home. And although more than half the Levant squadron was straggling by the end of August, all but one, or two at most, continued at sea until the third week of September. At least half the squadron was still afloat when the worst of the weather set in.

The squadron of hulks included some of the Armada's most unweatherly ships. Of the fighting ships, only the *Trinidad Valencera* was utterly inept in a head wind, though almost all the Levanters found progress in adverse conditions dangerous and slow. In the squadron of hulks weatherly ships were exceptional. The twelve or thirteen that got back to Spain in or by October, 1588, must for the most part have followed Pedro Coco

Calderón's advice and stuck rigidly to their sailing instructions, gaining enough Atlantic sea-room by means of the north-easterlies to cope with the struggle home through the latitudes of the westerlies. Even this course could not guarantee survival: the hulks suffered an exceptional rate of loss on the open sea, where at least three of them may have come to an end, victims, presumably, of the severity of the September storms. Relatively few tried their luck in Ireland, for the same reason: only six or, if the *Duquesa Santa Ana* is counted amongst them, seven. Only ships that could have some hope of making headway so as to return home afterwards were likely to yield to the temptation of an Irish refuge. Of the hulks that did go to Ireland, only one, known as the *Barca de Anzique*, is known to have escaped.

When all the great ships are added together, including the Levanters and hulks, the Armada's known losses can be seen to have totalled twenty-nine or thirty such craft, with at least fifty-four certainly surviving. Those crude figures seem horrifying, yet they mask some extenuating circumstances and some facts which are entirely to the Armada's credit. The Levanters and hulks were precisely the classes of shipping which

St Peter's Church, Tiverton, Devon, is decorated with extraordinarily rich carvings of about 1517, showing Mediterranean shipping of various kinds. The representations seem to betray the influence of manuscript marginalia or foreign woodcuts. The deck of one of the oared vessels, for instance, is manned by fanciful figures. The ships, however, are accurately handled by the sculptor and are of types well known on English coasts. Galley-borne trade probably helped to make the funds which paid for embellishments such as these.

Medina Sidonia had tried to dissuade the king from including. They were dispatched in full awareness that they would handle poorly in bad weather. yet they proved surprisingly robust, surviving the worst of the fight and all but the worst of the weather. The Atlantic-built fighting ships left little to be desired in qualities of durability. The three that fell to English guns were caught at a severe disadvantage; almost all of those, in all categories, that failed to cope with the adverse weather had some known and peculiar weakness, derived in the cases of many hulks and Levanters from their overall build, and, in the case of the galleasses and certain Atlantic ships, particular defects. Of the vessels that were able to adhere to the fleet's sailing instructions, and to follow the route appointed by Medina Sidonia, no single craft of any class of shipping is known to have come to grief.

It was therefore on the fate of the stragglers—the ships that got separated from the main fleet, discarded their sailing instructions, and, in most cases turned aside to Ireland—that the outcome of the Armada hinged. Something like two-thirds of the stragglers ended badly—most of them foundered, abandoned, or wrecked on or off the fatal Irish coast. There lay the difference between an acceptable rate of attrition on the one hand and disaster on the other. The deficiencies of Spanish strategy—above all, as I have suggested, the failure to provide a northern port of refuge—bear some responsibility for the Armada's failure; the English made a contribution of sorts to their own salvation; the weather did much of the rest; but the Armada would still have been reckoned a remarkably successful venture but for the work of the Irish siren. The fate of the stragglers therefore demands close scrutiny.

Nine

————— ⟫✖︎⟪ —————

Ariel's Victims

THROUGHOUT the voyage of the Armada the weather had been an almost unremitting foe. It was, with few exceptions, extraordinarily bad and became progressively worse. It was an appreciably more persistent and immeasurably more effective opponent than the English. The bad weather started even before the Armada set sail, when it was already unduly squally and cold—cold enough for Medina Sidonia to hope shipboard infections would be checked. Stalled in Lisbon by head winds, then dispersed to northern Spanish ports by a storm severe enough to strip the mainmast from the *Santa María de la Rosa* and swamp Recalde's *Concepción Mayor*, the Armada was further delayed, and its rations wasted, by an unprecedentedly rough July. Howard's complaints of the storms and of the 'extreme foul' weather leave no doubt of the reality of adverse conditions. It is sometimes supposed that the weather was bad only in the eyes of the beholders and that the Spanish experienced difficulty merely because they had unweatherly ships and unseamanlike practices. Yet Howard's words to Walsingham of 22 July—'I know not what weather you have had there, but there never was any such summer seen here on the sea'—are objective testimony enough. The weather in the Channel was bad enough to keep Parma's only galley penned in the Texelstroom and the Dutch flyboats in Flushing. 'The like', according to one of Walsingham's agents, 'hath not been seen by any at this time of the year.' When the Armada finally got out of Corunna, the galleys were almost at once parted by heavy seas. On 27 July, by Medina Sidonia's account, 'the sea was so heavy that all the sailors agreed that they had never seen its equal in July. Not only did the waves mount to the skies, but some seas broke clean over the ships ... It was the most cruel sight ever seen.'[1]

Good weather, it must be remembered, was essential for Spanish success. It was the least of the miracles they expected from God. They needed good weather to preserve their fragile Mediterranean shipping, calm seas to maximize the effectiveness of their galleasses, clement conditions for Parma's barges, a moderate swell to compensate for the inexperience of their gunners; and, above all, an easy, speedy voyage to help keep them safe despite the lack of a northern port of refuge. Yet the only sustained period of good weather they enjoyed was during their perilous ride up the Channel, when the following wind actually favoured the English more by guaranteeing the pursuers' command of the weather-gauge. There were only two days on which abnormal turns in the weather helped them—the first, as we have seen, on 9 August, when the 'Catholic wind' blew them off the Flanders banks, the other nearly a fortnight later when the Armada was able to shorten its route round Scotland. For the rest, the uniform adversity of nature or providence was a challenge to endurance and a trial to faith.

It is against this background that Protestant propaganda which interpreted the extraordinary weather as evidence of divine intervention has to be understood. Dutch sources were more insistent on the point than English ones, perhaps because the Dutch scrupled less to deprive their allies of the glory of a share in the outcome. It was Dutch commemorative medals which praised the 'quaking of the vasty deep', the dispersive power of 'Jehova's breath', and an England 'strong not in might but in righteousness'. English language tended to be more restrained. Indeed, the English medal which depicts the invulnerable island in the form of a green bay tree must have been conceived without awareness of the image's unfortunate scriptural connotations as the symbol of flourishing evil. Other English sources, however, were quick to exploit the weather with due humility. Sir Richard Bingham was loath to deprive Elizabeth of the satisfaction of a reputed victory over the Armada ('defeated first by Your Majesty's navy in the Narrow Seas') but acknowledged that the enemy had been 'sithence overthrown by the wonderful handiwork of Almighty God'. That 'the heavens did fight for us' was a source of assurance to Thomas Nash's *Piers Penniless* and doubtless to many of his countrymen. The queen herself, in her own lines written in thanksgiving for her deliverance, is frank about her debt to the weather and presumptuous about the source of it:

> He made the winds and waters rise
> To scatter all mine enemies.

Stripped of their providential language and imputations of divine bias, such assertions were entirely consistent with the facts. Philip II was said to have consoled himself that he had sent the Armada 'against men, not against the sea and the weather'. He may not have said so, but the attribution shows that the decisive impact of the weather was admitted in Spain, despite its propaganda value to the enemy. To admit the role of the weather was effectively to perceive the hand of God, for winds were such obvious vectors of divine influence in the world. Weather of such peculiar and intense and selective malevolence as that of 1588 could hardly be explained as wholly natural. What for Protestants was a vindication was for Catholics a temptation; the two sides differed chiefly in seeing divine intervention as specific to themselves—Protestants as the reward of righteousness, Catholics as a punishment for sin. To the one side, the weather seemed 'ordained', to the other merely 'allowed'. 'In the storms through which the Armada sailed,' wrote Philip to his bishops on 13 October 1588, when almost the full extent of the débâcle was known, 'it might have suffered a worse fate, and that its ill fortune was no greater must be credited to the prayers for its good success so devoutly and continually offered.' This was as close as anyone came on the Catholic side to relegating God to His heaven and the weather to the strictly natural sphere.[2]

The weather's ravages should not be seen as entirely fortuitous. Their effect depended on what was perhaps the main weakness of the Spanish tactical plan: the failure to provide a safe haven of refuge for the Armada in northern waters. The want of this elementary amenity had been shown in the Armada's desperate recourse to a dangerous anchorage at Calais and then again by the inevitable decision to 'obey the wind' and sail up the North Sea on 10 August. This was why the Armada came to be at the mercy of the September weather that destroyed so much of it, when it ought to have been in snug quarters not far from Parma's army. As the only ports that might have served were on the south coast of England, where Medina Sidonia had been forbidden to land, or in the northern Netherlands, which Parma had been forbidden to conquer, the decision-makers in Madrid and the Escorial must be held responsible for this deficiency, and share the weather's blame for the wreck or disappearance of most of the Armada's stragglers. Some of these disappeared without trace. Of others only the site of their end is known or suspected. But the survival of some documents and the exploration of some wrecks make it possible to reconstruct in outline the fate of about half a dozen major groups, including the most important of the lost ships.

Most of them ended their voyage grounded or shattered on the west coast of Ireland. Medina Sidonia's sailing instructions could not have been more explicit about the dangers of that coast: 'take great heed lest you fall upon the island of Ireland, for fear of the harm that may happen to you upon that coast'. But, while this warning was to be tragically justified, there were clearly some captains who were reluctant to accept it without demur. The issue was twice debated aboard the flagship in the course of the voyage: first, in the North Sea, when the sailing instructions were

Painted by George Gower (*fl.* 1570–96), Elizabeth is more like a hieratic emblem than a creature of flesh and blood. The crown imperial which rests on the table beside her topped the canopy beneath which she was borne in triumph into London to celebrate England's deliverance from the Armada on 24 November 1588. Crucial scenes in the Armada's undoing—procured respectively by human and divine agency—are glimpsed through the windows: the fireship attack on the left, and the Atlantic storms, brilliantly evoked on the right. Note the aquatic caryatid, who resembles a ship's prow carving, at bottom right.

issued; then, if Pedro Coco Calderón's account can be accepted, off the north-east coast of Scotland. It must be said that Pedro Coco's evidence is doubly partisan: he wrote partly to join the general denigration of Diego Flores de Valdés, accusing him of supporting the insubordinate and—as it turned out—frequently fatal plan of seeking Irish havens; and, in part, to glorify himself, claiming on his own behalf an implausible degree of navigational expertise and arrogating to himself the role of apologist for his commander's orders. There is an unpleasant glibness about his account of his ship's fortitude in beating out to sea, with all hands on the bowlines, to avoid the Irish coast and 'it is believed that the rest of the Armada will have done the same: if not, they will certainly have lost some ships'. The scene, however, of a tense debate aboard the flagship, with rations running out and an indefeasible head wind before them, seems well imagined: in those circumstances, a dash for Ireland must have seemed worth the risk. Moreover, there were pilots aboard the Armada, who, in better weather, might have been capable of finding an Irish anchorage—some, perhaps, because they were Irish themselves, others, like Juan Martínez de Recalde, because they had been there before. Recalde, as we shall see, was capable of steering a flotilla in and out of Blasket Sound in safety in heavy seas— no mean feat in the best of weather. Some Spaniards captured at Tralee claimed to have friends in Waterford. A few ships got independently into Killibegs and Blacksod Bays in safety; one of them, the *Duquesa Santa Ana*, found a secure place to lay to; and two of them—the *Girona* and the *Nuestra Señora de Begoña*—actually made it out to sea again. This is not to say that such operations could have been successfully attempted on a larger scale, with pilots whose ignorance of the coast was otherwise almost total. Few of the ships that made for Ireland reached land safely and fewer by far escaped to tell the tale. Yet there were such ships and it is a tribute to Spanish discipline that crews, who might have saved themselves by a diversion to Ireland, stayed with the main fleet for so long, or, after separation, continued, in most cases, to follow it so doggedly. Even some of the earliest and most laggard stragglers—the *Gran Grifón*, the *Rata Encoronada*, the *Trinidad Valencera*—who can have had little hope of rejoining the flagship, struggled on until the unendurably bad weather of the second and third weeks of September. The only Spanish ships that tried their fortune in Ireland were forced there by necessity or despair.

The most important and most successful group centred on the *San Juan* of Juan Martínez de Recalde. He had been shepherding a little fleet of stragglers, at first thirteen to fifteen strong—maintaining tenuous contact with the main fleet—since the storms of the third week of August. By about the end of the month, or early in the next, he lost contact with the flagship and led off a raggle-taggle company of twenty-seven sail—by the only account we have—in a mood of desperation, but as yet with no apparent aim of seeking refuge in Ireland. It was only the severe storms of the second week of September, sundering the ships from one another so completely that they seem to have lost all hope of reforming, that compelled Recalde to turn aside. Accompanied now only by a single pinnace, he made for the Blaskets, of which he seems to have retained a vivid pilot's memory from his voyage in support of the papal expeditionary force at Smerwick in 1580. What happened on his arrival and in the dangerous days that followed is recounted in the diary of a squadron purser, Marcos de Aramburu, aboard the galleon *San Juan*, vice-flagship of the squadron of Castile. His ship had been having sailing difficulties since losing her foremast on 26 August. Two days later she had lost touch with the main fleet and by the 12 September she was wandering alone, making scant headway and edged ever closer to the Irish coast by gales. It seems only to have been the chance meeting with Recalde on 15 September that definitely resolved the course of the *San Juan* in favour of Ireland. There is no evidence that anyone aboard her had more than a chartsman's knowledge of the Irish coast, and she would have found it impossible to find her way safely into Blasket Sound had she not been able to thread through the perilous entrance in the exact path of Recalde's vessel:

We followed her with a crosswind blowing, knowing nothing of the land, despairing of any remedy, and we saw how, being able to double one of the islands by way of another land that he saw in front of him, he made his approach from the east. We stood to windward of his ship and continued to follow him, thinking he had some knowledge.

It was an impressive piece of pilotage, which Recalde had risen from his sick-bed to execute. The two great ships anchored tentatively within, in a heavy sea, with makeshift anchors always threatening to drag, hulls actually clashing with an ominous sound. At first hemmed in by the gales, then tarrying to take on water, they had been at anchor for a week when a ship appeared which they recognized at once. The *Santa María de la Rosa*, at 26 guns the second most heavily armed vessel of the Guipúzcoan squadron, seems to have approached through the northern entrance to the Sound, rather than from the east, as favoured by Recalde. In normal conditions, the northern channel was preferable, but fatal in a crosswind and a heaving sea. She came in firing off signal guns as a sign of distress, perhaps already awash from storm damage, perhaps, as the only account to have survived from one of the ship's company suggests, getting holed on the rocks on the way. Aramburu watched her:

She was wearing all her sails in tatters, except for the foresail. She found the bottom with a single anchor, for that was all she had, and on the flood tide that was coming in from the south-east, which bore her up, she remained until the second hour, when the tide began to ebb. And as the tide turned she began to drag to within two splicings of us, and we dragged with her, and the next moment we saw her start to go down, trying to hoist her foresail and then she sank with all hands, not a man of them escaping, which was a very amazing and frightening thing to see.

In fact, unobserved by Aramburu, an Italian cabin boy survived by clinging to flotsam. He was washed ashore, captured by the English, interrogated, probably under torture, for information he did not possess, and put to death. He had time, however, to spin a yarn that seems to have deceived his tormentors into thinking that the *Rosa* was a major ship, full of notables including Oquendo and the Prince of Ascoli; the substance of his interrogation found its way into a popular English propaganda pamphlet, and thence into a corrupt historical tradition.[3]

The excavation of the site of the wreck by Sidney Wignall's team in 1968 revealed exactly how the disaster happened. The ship had virtually snapped in half on the rock known as Stromboli Reef when her anchor dragged. The bow section plunged to the bottom, with her mast and the bulk of the ballast, most of which was stowed for'ard to stabilize the forecastle and the eccentrically placed mainmast. The ballast from the stern section largely spilled out too, leaving the rest of the hull to drift off like a cockleshell on the tide.[4]

At four o'clock on the same day, one of the auxiliaries of the Castilian squadron, the *San Juan* of Fernando de Hora (whose name is also recorded, less plausibly, as Fernán Dome) limped into the Sound without a mainmast. This was a small, light, trim but battle-worthy vessel, sometimes mistaken for a galleon, but not classified as such officially. She might have been an early example of a *gallizabra*—a new, experimental type of vessel built on the lines of a galleon, heavily gunned for her size, and vying with a large pinnace for speed. According to Pedro Coco Calderón, she was, in her prime, one of the two fastest great ships in the Armada. 'As she entered her foresail fell to pieces. She found the bottom and hove to. The weather was so bad we could get no message from her nor help her.' She was too badly storm-damaged to save. The other two *San Juans* took off her company the next day and sailed away for Spain on the 23rd, probably with a second pinnace in the flotilla too. The formation was soon broken up by more storms. Aramburu's ship sighted Spain on 8 October, but the continuing perversity of the weather kept her off the shore, repeatedly buffeted back into the Atlantic. At last she made port at Santander on the 14th. Recalde also reached home, dying of exhaustion. His achievement in finding and holding a safe anchorage in conditions of the utmost adversity, and saving four ships and the crew of a fifth, was redoubtable.

Only one other group of stragglers to Ireland encountered comparable success. On 12 September a flotilla led by the Ragusan armed merchantman *Anunciada* (perhaps one of the Levanters reported lagging since the North Sea) found a good anchorage in the mouth of the Shannon,

probably in Scattery Roads. Evidence of the nature of this group contains
conflicts which are hard to resolve. According to a crewman's deposition,
recorded later in Spain and Dubrovnik, the *Anunciada* had been extricated
with difficulty from the battle of 8 August and accompanied thereafter by
five pinnaces, who by implication all put into the Shannon together. Sir
John Popham, however, writing from Cork on 20 September to report the
presence of this group and others to Lord Burghley, had the impression
that there were no more than three pinnaces with the *Anunciada*. A
reasonable hypothesis might be that the *Anunciada* was accompanied at
this stage by three pinnaces, two having left her or been separated in the
interim, and two larger ships. On the 13th the large Baltic hulk the *Barca
de Anzique* (that is, Gdańsk) arrived, bringing the group up to the total of
seven reported by Popham, and probably roughly corresponding to the
categories he identified: two large, two medium, and three small ships.
Among the pinnaces, it seems likely from Ragusan traditions that one
was Ragusan-built, known in Dubrovnik as the *Sveti Jerolim* and therefore
corresponding to the *San Jerónimo* that appears among the squadron of
pinnaces in the Armada's battle order. The group spent nearly a week
repairing the storm-damaged hulk and stripping the battle-scarred fighting
ship. The *Anunciada* carried 24 guns, 80 crew, and a complement of 186
troops, as well as some unperished stores, to redistribute among the other
ships. On the 19th she was fired where she lay. An adverse gale kept the
flotilla in the estuary that night, but on the 20th the wind abated and

Memorial to members of the
Iveglia family who served with
the Armada in a Franciscan
church in the hinterland of
Dubrovnik, dated 1590. The
combination of a *memento mori*
with an inscription celebrating
the dead brothers' noble
ancestry might belong to any
monument of the Latin West at
the time.

they sailed away without further mishap. Although the *San Jerónimo* is missing from the Spanish lists of ships returned, the *Anunciada*'s captain, Stefan Olisti Tasovčic, who was Petar Iveglia's nephew, shipped aboard her and got home safely to serve the Catholic cause for many years.

The very wind that detained these craft brought two other ships to grief nearby, both wrecked on 20 September. It was probably the Guipúzcoan *San Esteban* that was lost in Doonbeg Bay with most hands drowned. A second ship of debated identity broke up suddenly on a reef by Mutton Island off the castle of Tromra at Kilmurry-Ibrickane. The sea was so heavy that only four survivors managed to swim the comparatively short distance to safety. It is possible that the Mutton Island wreck was the *San Marcos*. Certainly, at some point in the voyage, on that great Portuguese galleon that had fought so resolutely alongside the flagship and suffered so terribly in the battle of 8 August, her wounds finally began to tell. The flagship lost touch with her between 21 and 24 August. If, however, her loss is to be ascribed to English gunnery, it must be said that the effect was very belated: it had taken more than a month of recurrent battering from the weather and the lash of a hurricane's tail to complete the enemy's

work. In a persistent tradition, moreover, she was believed to have been scuttled, not wrecked. The case for her identification with the Mutton Island ship is imperfect in a crucial respect, based on the identity of a distinguished officer, Don Felipe de Córdoba, who had fought on the *San Marcos* and was reported captured by the English. However, as he was also recorded, by several independent witnesses, as among those killed on deck in the fighting of 8 August, it seems imprudent to rely on evidence connected with him. It may be preferable, as a working hypothesis, to adhere to the tradition that the *San Marcos* was scuttled and burned, probably in some other, unknown, location. The evidence for her disappearance is not entirely conclusive, and she may even have survived to return to Spain to be broken up for her timber; but, if so, I have found no record of her return and therefore continue to count her among ships lost.[5]

Of the remaining groups, the story is almost uniformly melancholy. The first group to fall away was that led by the large fighting hulk, the *Gran Grifón*, from about 18 August. The *Gran Grifón* was a highly esteemed vessel, the flagship of the hulks, appropriated for his squadron by Recalde during the fighting. Her captain, Juan Gómez de Medina, had been transferred from a galleon and was a man of enough consequence to be cited by name with only five others in Aramburu's list of 'principal men' lost or delayed on the voyage home. The ship occupied a conspicuous place in the battle line, bore the brunt of the Channel sparring of 3 August, and performed honourably on 8 August. The diary which has survived from on board, however, suggests that she was a poor sailor who found it hard to cope with a severe head wind. In her company was the *Trinidad Valencera*, the third in rank of the Levant squadron and fourth largest ship in the fleet by tonnage. Commanded by one of the Armada's most distinguished soldiers, Don Alonso de Luzón, she had played an important role on 8 August, going to the relief of the Armada's hard-pressed weather wing, together with the *San Juan de Sicilia*. She must have taken her share of the hard pounding inflicted by English ships—and those of Seymour's squadron in particular—in that quarter of the battle. Nor were her sailing qualities good at the best of times. Together with the *Gran Grifón* she had been obliged to run north by the storm that drove most of the Armada into Corunna in June and on that occasion was one of the last ships to return to the fleet. The other two ships that kept her and the *Gran Grifón* company in late August were supply vessels, the *Castillo Negro* and the German hulk known as the *Barca de Amburg*—unweatherly waddlers who had little hope of making headway in the conditions through which the Armada had to sail. The last named was abandoned and her company divided among the rest. The *Castillo* went unrecorded after the beginning of September and may have foundered in the open sea during one of the terrible storms of that month. The two main ships of this dogged little flotilla were separated from one another on 2 September.[6]

The *Trinidad Valencera* was the first to give up the struggle against the weather. On 12 September, faced with a storm from the south-west, her officers felt they had no choice but to run with the wind and trust to luck or providence. They reached Kinnagoe Bay on 14 September and found an anchorage by good fortune on an inhospitable shore, of which they had no adequate knowledge. The ship was sinking. They had only one boat and no apparent hope of getting a ship's company near four hundred strong ashore in time. But the natives, who soon thronged the shore, were disaffected with the English and prepared at a price, to help a foreign friend. Even with the curraghs of the O'Dohertys to ferry them, it still took two days to bring the Spaniards off, and when the ship at last went down on the 16th there were still forty hands aboard. The *Valencera* also took with her some of the Armada's finest guns. She carried four of the fleet's great siege battery, destined for the land war in England, including a pair of fifty-pounder cannon. Short-barrelled and lightly metalled, built expressly for siege warfare by Rémy de Halut, one of the foremost gun founders of his day, at Malines in 1556, they were retrieved from the bottom of the bay by Sidney Wignall's divers in 1970.

Luzón was now faced with a difficult dilemma. The O'Dohertys were by no means uniformly sympathetic to his cause. Their chief, Sir John O'Doherty, was untrustworthy and, indeed, seems to have been intending to betray the Spaniards. Many members of the clan, while hating the English, recognized the reality of English power and the need to compromise with it. Luzón toyed with the idea of seizing an O'Doherty castle; but little long-term security could be gained from such a move. There were thus likely to be few helpers and no comfortable quarters for his men in the vicinity of the wreck. Though the nearest sure sympathizers—the McDonnells and O'Cahans of the north-east corner of Ulster—were a long way off by land, around Lough Foyle, there were no large English forces known to be in the way, and Don Alonso had an élite force under his command. He set out on the long march across country.

There was never any chance that he would complete the journey. The English governor was already aware of his presence and a force of pro-English cavalry, led by clients of Hugh O'Neill, Earl of Tyrone, who at that time was 'affected' to the English cause, was on its way to monitor his movements. 'We be 150 men,' wrote the officers in charge of it, 'and will God willing be doing with the Spaniards ... though we are in doubt

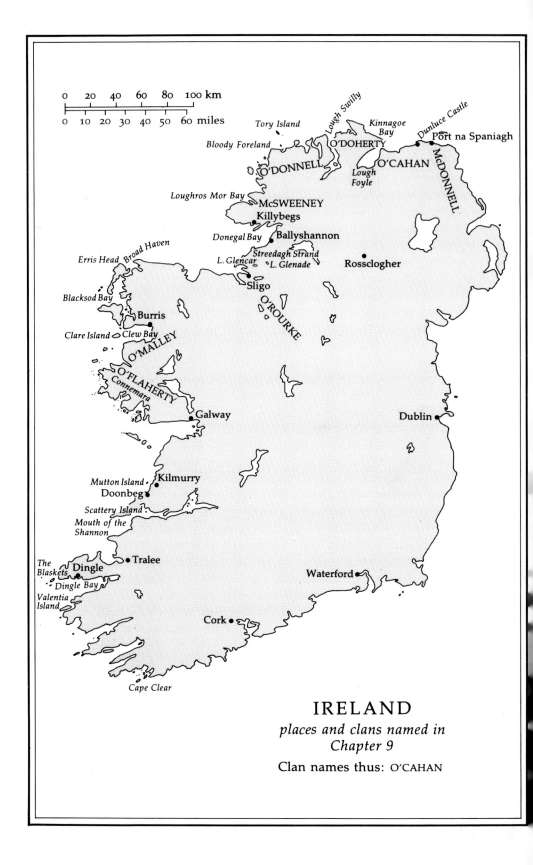

Scale:

0 20 40 60 80 100 km

0 10 20 30 40 50 60 miles

Tory Island

Lough Swilly

Kinnagoe Bay

Dunluce Castle

Bloody Foreland

O'DOHERTY

Port na Spaniagh

O'DONNELL

O'CAHAN

McDONNELL

Lough Foyle

Loughros Mor Bay

McSWEENEY

Killybegs

Donegal Bay

Ballyshannon

Streedagh Strand

L. Glencar

L. Glenade

Rossclogher

Erris Head

Broad Haven

Sligo

Blacksod Bay

O'ROURKE

Burris

Clare Island

Clew Bay

O'MALLEY

O'FLAHERTY

Connemara

Galway

Dublin

Mutton Island

Kilmurry

Doonbeg

Scattery Island

Mouth of the Shannon

Tralee

The Blaskets

Dingle

Waterford

Dingle Bay

Valentia Island

Cork

Cape Clear

IRELAND

places and clans named in
Chapter 9

Clan names thus: O'CAHAN

whether the countryside be true to us or not.' When he encountered this party, Luzón's first inclination was to offer battle. He rejected the enemy's calls for surrender without guarantees of safe conduct; he was not, however, entirely insensible to peaceful overtures. His force of infantry could hold off but not decisively defeat the horsemen. Their march could make but slow progress under enemy harrassment; and he faced a long struggle of attrition which might at any moment be transformed into a rout by the arrival of large enemy reinforcements. On the second day of the confrontation he had a parley with Major Kelly, who seemed to be the figure of greatest authority on the enemy side. Assuming he was dealing with honourable men, Luzón accepted the promise of a safe conduct to Dublin and ordered his men to lay down their arms. The promise seems to have been supported by the presence of a prominent member of the O'Donnell clan, whose renown as an opponent of English rule appeared to Luzón a guarantee of integrity. The Spaniard was unaware that the clan was divided and duplicitous. Sir Hugh O'Donnell's son was a hostage and his partisans would stick at nothing to secure his safety or liberty, while his enemies, the clients of Sir Niall O'Donnell, would do anything to keep him confined. Luzón also had a touching faith in the efficacy of 'the queen's ensigns' as indicators of the probity of those who bore them. The agreement he made with Kelly was that every private soldier would be allowed to keep one suit of apparel and every gentleman two. Yet as soon as they had surrendered their firearms 'and some few pikes . . . the soldiers and savage people spoiled them of all they had'. This was the least of the offences committed against the terms of surrender. The following morning, after about forty officers had been separated from the rest, the common soldiers were put to the slaughter in a roadside field, by a volley of musketry and a cavalry charge, discharged and unleashed against naked, unarmed men. The only reflection that can be offered in mitigation of this massacre is that it was inefficiently carried out. Some hundred men escaped, to be helped by the Bishop of Down who gradually smuggled them out to Scotland via the McDonnells and O'Cahans of Dungiven. At Dunluce Castle, Sorley Boy McDonnell granted lodging and transport to at least eighty of them with all the magnanimity of a noble 'savage'. He 'would not sell the blood of a Christian man' and ordered twenty masses a day to be offered for their safety. Meanwhile, on the other side of the Foyle, among the survivors of the first massacre, only the two most promising candidates for ransom—Luzón himself and Don Rodrigo Laso, who had shipped with five fighting retainers of his own—were spared. Eventually they passed into captivity in England.[7]

The rules of war have rarely been applied in Ireland: in the conditions of 1588, with rumours rife of Spanish ships upon the coast and hun-dreds—by some reports, thousands—of Spanish troops at the service of rebels, the commanders of English garrisons were under a terrible strain. The difficulty of accommodating prisoners securely, the threat posed by large numbers of rebels already affected by sympathy and united by religion to Spain, and the habitually brutal conduct of war in Ireland all provide a context in which the massacres of the Spaniards become intelligible. Despite the opportunities for ransom, ruthlessness was untem-pered even by greed. Sir Richard Bingham in Connacht thought wistfully of the fortune he had forfeited by complying with the Lord Deputy's orders to liquidate prisoners, almost to a man, even when 'the heat and fury of justice was passed'. Lord Deputy Fitzwilliam justified his merciless policy with a notorious and gruesome joke: since God had seen fit to cast the Spaniards on the shore, many of them already dead, Fitzwilliam would be his deputy 'for the despatching of the rest'.

The *Gran Grifón*, meanwhile, had endured longer and fared better. She went on trying to make her way south until the fateful storms of 17 September: though she must have been a long way behind the main fleet, the expiring hurricane still managed to catch her. Her seams were parting and she was letting water badly, with the soldiers working the pumps non-stop. She ran north for three days, then spent three days more trying to resume her former course. By the 24th, however, even this remarkable display of determination was beginning to falter and the officers decided to try to find a Scottish haven. On the 25th, our diarist was willing to settle for 'the closest land'. The 26th and 27th were spent struggling against adverse conditions in an attempt to gain an island anchorage, probably at North Ronaldsay:

We wanted to gain the protection of the island, for we were afraid of the sea and the heavy weather. We did all we could, struggling to make headway against wind and sea for some two or three hours until we saw that it was impossible . . . All the time the sea was battering us with great blows and truly we thought that it would put an end to our lives. And so the men made their peace with God as they ought and each as well as he could, to be ready for the final journey that we were sure we saw ahead. As any further exertion with the hulk would only have ended our lives even sooner we decided to give up the objective we desired so much to endure whatever the hulk endured. And the Lord was pleased to spare our lives and those of the soldiers, who were fainting and had no strength to man the pumps, seeing how hopeless their efforts were, for the water kept rising all the time in the ship until it was thirteen hands' breadths over the carling, as it is called. And neither pumps nor buckets would suffice to lessen the water by a single finger's breadth. And so we were in despair, every man calling on the

Virgin Mary to be our intercessor in that most bitter journey, not neglecting withal our labours at the pumps and with the buckets, staring at the land with such eyes and hearts as the reader will be able to imagine, and please God he may be able to guess only an infinitesimal part of what it was like, for, after all, suffering will always be worse for the man who undergoes it than for him who observes it at a safe distance. In the end, despairing of any remedy save from God, Who has never spurned those who call upon Him, and with the help of His mother, who never tired of praying to her Son, at about the second or third hour of the afternoon, with a suddenness that surpassed all suddenness we saw an island ahead. It was recognised as Fair Isle. We arrived as the sun was setting, thoroughly consoled, although we realised there had to be more suffering ahead, for as an alternative to drinking salt water anything seems preferable.

The further suffering began almost at once: the Spaniards' attempt to beach their ship came to grief when, carried perhaps by a strong tide, they were wedged in the Strombshellier cave, under a cliff, with their masts leaning on the rock face. They had to disembark by climbing the rigging up the masts and over the yards. They managed to take off their treasure, but 'without being able to save anything to eat'.

At that point, the ship's company had survived the voyage well. The *Gran Grifón* had started with an official muster of 243 soldiers and 43 sailors. Of the former, thirteen were dead—most of them probably casualties of the battle of 2 August, when the ship had been caught in the thick of the fight. Forty others had been taken aboard at the abandonment of the *Barca de Amburg*. The diarist reports that 300 men landed on Fair Isle, and it is likely that he is rounding down rather than up, as he also claims that fifty died of hunger during the long stay on the ill-provisioned island, whereas we know from another source that 260 were eventually rescued. That rescue was long deferred. Not until 27 October did the marooned Spaniards get a message to the Scottish mainland. Not until 6 December were they able to disembark at Anstruther on the Firth of Forth, 'for the most part young beardless men, silly, traunchled and hungered, to the which kail, porridge and fish was given'. In the interim, they had lived on what little food could be spared for cash by the seventeen poor families that lived on Fair Isle. They found the natives unimpressive: 'wild folk', our diarist calls them—perhaps with a conscious echo of St Paul, marooned among the 'barbarous' but generous Maltese—'very dirty ... neither Christians nor yet altogether heretics'. The Spaniards, however, took comfort from the inhabitants' evident distaste for the Calvinist missionaries who visited from the mainland, and from the extraordinarily hospitable reception the people were willing to give to this large band of grim warriors from the sea. Despite the extreme necessity in which they found themselves, and the deaths attributed by the diarist to hunger, the Spaniards maintained an attitude of perfect discipline and, with the natives, of mutual respect. The island folk seem willingly to have spared what they could and their guests scrupulously to have paid for it. When they got to the mainland, the fugitives showed the same diplomatic qualities; Gómez de Medina impressed the dour officials with whom he dealt by his courtesy and deference. Perhaps as a result, his men continued to be well treated and their return to Spain expedited. At home, Gómez de Medina became a spokesman for Scotland. He advocated the alliance proposed to him by Scots Catholics and crypto-Catholics who fêted him in Edinburgh; he procured the release of an impounded ship of Anstruther, and extolled the virtues of the Shetlanders.[8]

Not long after the separation of the *Gran Grifón*'s group from the main fleet, Don Alonso de Leiva found himself cut off from the Armada in his Genoese fighting carrack, the *Rata Encoronada*. Every day from 21 August onwards the question, 'Where is Don Alonso de Leiva?' was asked disconsolately around the fleet, but no one reported seeing him and the best that was hoped was that he had run for Iceland or the Faroes. In fact, he was still pursuing a homeward course, but his unwieldy and battle-scarred Levanter was 'riddled with small shot, pierced in places [and] leaking badly', in words attributed by Vanegas to de Leiva himself. She made even less headway than most of her sister ships. She was reunited with other Armada vessels by a curious mischance. On 10 September, de Leiva felt obliged to give up the attempt to get home and to make for an Irish port. It was not, however, until the 17th that he found his way into Blacksod Bay. This was, of course, precisely the time of the Atlantic tempests that scattered so much of the Armada, driving back to Ireland many vessels that had almost reached Spain, and the *Rata* began to encounter them at once. In the bay, or at the approach to it, she was joined by the *Duquesa Santa Ana*, the Andalusian hulk, which, whatever the condition of the *Rata*, was certainly seaworthy. Both ships found an anchorage. The *Rata* was stripped and fired, perhaps only after dragging her anchor and drifting aground, and 'Don Alonso and all his company were received into the hulk of St Anne with all the goods they had in the ship of any value, as plate, apparel, money, jewels, weapons and armour'. The covetous eyes that recorded this scene belonged to an Irish sailor aboard the *Santa Ana*. The overburdened hulk was detained in Blacksod Bay by adverse winds until the 27th, but then set sail, presumably bound for Scotland: the *Santa Ana*'s complement was 280 soldiers and 77 sailors. Don Alonso had brought some 400 men from the *Rata*. A short voyage was all that could be contemplated.

In the event, it was cut short by the wind and the loss of the last usable anchor. Probably on 29 September, the *Duquesa Santa Ana* went aground at Loughros More by Rossbeg. De Leiva was hurt in the wreck or the disembarkation and had to spend a while strapped in a litter, but otherwise the Spaniards again seem to have kept their company intact. At least the weather had now consigned them to a part of Ireland ruled by men uniformly hostile to the English. There could be no massacres in the country of the McSweeneys. De Leiva now received the inspiring news that three more Spanish ships had found refuge with the McSweeneys at Killybegs. Closer inspection revealed that two of the three were wrecked beyond repair. The third, however, the galleass *Girona*, needed only relatively modest repairs. Like all the galleasses, she had trouble with her rudder, but, if this could be replaced, she could be made seaworthy enough. The only insuperable difficulty would lie in transporting the companies of five vessels in a single ship, probably over 1,500 men— some 1,300 of whom seem actually to have been crammed aboard, space designed for about 550. The English could wish Don Alonso Godspeed. The news that a large Spanish force was lodged with some of the worst enemies of English Ireland in a highly defensible spot caused a fear bordering on panic in Galway, Dublin, and London, especially as the reports the English received greatly exaggerated the size of de Leiva's units. In the event, the Spaniards seem to have decided that their own best prospects lay in a dash for Scotland. The McSweeneys, though anxious for allies, could not feed them indefinitely; and insufficient stores could be gathered for the voyage to Spain. By 26 October, repairs to the *Girona* had been effected, and de Leiva's last voyage began. At first, he seemed set fair to escape. Then, on the 27th, the wind, which had frustrated all the Armada's efforts to get to Spain, veered round the north to thwart their voyage to Scotland. Some of the seamen had a last flicker of hope: the wind would 'take them back to Spain'. But it only blew them towards the rocks of Ulster. The circumstances of their final wreck suggest that their makeshift rudder had failed. the *Girona* splintered on the rocks and reefs of Lacada Point, at the spot still called Port na Spaniagh, where Robert Sténuit discovered its wreckage in 1967. Don Alonso and all the other gentlemen aboard were lost. There was only a handful of survivors, nine of whom found refuge with the hospitable Sorley Boy McDonnell at Dunluce Castle, where, with survivors of the *Trinidad Valencera*, they were shipped to Scotland and thence to home.[9]

There was little hope for the Spaniards wrecked to the north-west of Galway. In that port, the grip of the English was unpopular but unshakeable. The less accessible coasts close by, where the English hold was

Left. A gold salamander set with rubies, excavated by Robert Sténuit from the wreck of the *Girona*. Salamanders were a popular decorative motif particularly suitable for soldiers' talismans, as they lived in fire: this one, ironically, was to perish in another element. Note the webbed feet which make the fire-dwelling lizard reminiscent of his aquatic cognates.

Right. The rich débris of the *Girona*: one of a set of a dozen Roman emperors' heads, carved in lapis lazuli and set in gold and pearls. Taken together, these are the most intrinsically valuable find to have emerged from the Armada wrecks and may have been intended as a gift for an important English collaborator.

weaker, in Connemara, were doubly lethal: those Spaniards whom the rocks and cliffs spared could be finished off by the predatory inhabitants. At least one ship was wrecked in this area by 16 September, at Burris, Clew Bay. It seems likely that this was the Ragusan, the *San Nicolás Prodaneli*, which as we know (p. 15 above), was reported lost 'on the cliffs of Ireland'. The relatively early date of this wreck, at any rate, suggests a Levantine vessel. By 24 September—the evidence is insufficient to yield the exact dates—four more great ships appeared on neighbouring coasts. One escaped, losing only an abortive landing party, perhaps seventy strong, who fell into English hands. The rest were wrecked: the *Falcó Blanco Mayor*, a troop-carrying hulk, and a ship of Recalde's squadron, perhaps the small *San Juan* of 21 guns, both seem to have been trapped against the cliffs of Connemara, while the *Gran Grín*, Recalde's vice-flagship, ran aground on Clare Island. From the *Falcó* and *Gran Grín*, at least, large parties got away, with their leading officers, in pinnaces or big boats, but that was their last stroke of good luck. The men of the *Gran Grín* were imprisoned and later slain by the piratical O'Malleys of Clare Island. Don Luis de Córdoba and the survivors of the *Falcó Blanco Mayor* tried to barter peacefully for supplies in Galway but were overwhelmed. From all these wrecks together, over three hundred prisoners who fell into English hands were put to death by the English garrison at Galway. Only Don Luis himself—evidently a gentleman of quality despite his brave protestations of poverty—is known to have been spared for ransom and eventually to have been returned to Spain. It seems sensible to suppose that the *Concepción*, *Falcó*, and *Gran Grín* may have formed part of Recalde's little fleet, scattered by the storms of the second week of September. The *San Nicolás* may have formed part of the same group, but, because of her poorer sailing qualities, got detached slightly earlier than the rest; alternatively, she may have been a lone struggler for some time, like some of her sister Levanters, and, like the *Anunciada*, may have been one of the three big Levanters seen to fall away in the North Sea.[10]

The last group of Armada ships to be wrecked in Ireland was pounded to smithereens on the beach of Streedagh Strand, near Grage, to the north of Sligo, on 25 September 1588. their fate is chronicled in the remarkably graphic letter written from Antwerp on 4 October 1589 by Francisco de Cuellar, announcing his escape and recounting his adventures. At the time of the disaster he was aboard the *Lavia*, a big Levanter, a privileged prisoner after the disciplinary imbroglio, from which he had narrowly escaped with his life, in the North Sea (above, p. 206). His account has to be read with caution. He was writing with emotion recollected in tranquillity, which must have heightened, and may have distorted, his memories. His letter, moreover, was characterized by a peculiar moral bias. The narrator was its hero, whose innocence was sometimes explicitly stressed—as in the account of the disciplinary process in the North Sea—sometimes symbolically expressed, as in his frequent experience of abject poverty and literal nakedness. His fate—which was to be consigned in turn to each of a series of extreme misfortunes from which he emerged only on the very brink of death—was contrasted with that of rich, noble, and powerful shipmates who all died horribly and often ironically. His epic journey to safety was trekked mainly through a topsy-turvy land in which savages are charitable Christians, 'a kingdom in which there is neither justice nor law, and everyone does what he pleases'. When he was rescued—as he often was—from apparently certain death, it was almost always by clerks, young boys, or women. And attractive young women intervened in the process with almost lubricious frequency. Yet at the same time, Cuellar is a master of self-mockery, with flashes of Pooterish humour. In short, the letter would be hailed as a masterpiece of literary contrivance, were its basic framework not verifiably true. It is hard to resist the feeling that the author must have modified the narrative to give it universal significance and to make of it a mixture of fable, satire, parable, epic, and cautionary tale.[11]

According to Cuellar, the *Lavia* was accompanied by two other very large ships 'to afford us aid if they could'. From internal evidence, one of these can be tentatively identified as the galleon *San Juan Menor* of the squadron of Castile, the only Castilian galleon to be lost in the course of the Armada campaign and voyage: in at least one Spanish list she is shown as returned home; but the lists are inconsistent and unreliable. Her senior officer, Don Diego Enríquez, had shipped as an adventurer, but seems to have assumed command at an early stage of the voyage. Medina Sidonia appointed him to take over the squadron of Andalusia when Don Pedro de Valdés was lost, and he seems to have taken his galleon with him. Pedro Coco Calderón praised his valour and the duke his seamanship. In the battle of 8 August her rigging was so shot up that all her sails had to be replaced: this seems likely to have increased her seaworthiness, but perhaps by late September other damage was taking its toll. Cuellar gives the impression that the group was lagging well behind the Armada, though he does not say why, save to imply that the *Lavia* was in need of succour, having been 'opened' by the force of the storms. The ships seem to have been further victims of the typhoon which raged from 17 September. On the 20th, despairing of clearing Ireland 'on account of the severe storm that arose ahead', the flotilla's commander, Don Diego Enríquez, allowed his ships to be blown on to the coast and was obliged to anchor in an unsheltered position in Sligo Bay. There they remained for four days, unable to find a safe anchorage or return to the open sea. 'On the fifth day,' Cuellar continues,

there arose so great a storm on our beam with a sea up to the heavens, so that the cables could not hold nor the sails serve us, and we were driven aground with all three ships before a beach covered with very fine sand, shut in on one side and the other by great rocks. No such thing was ever seen, for within the space of an hour all three ships were broken in pieces, so that fewer than three hundred men escaped, and more than a thousand were drowned, among them many persons of importance—captains, gentlemen and other officials.

Cuellar's numbers must be grossly exaggerated. His insistence on the 'very large' dimensions of the three ships calls to mind the impression made a few months later on the Lord Deputy of Ireland, who rode along the strand, marvelling at the

great store of the timber of wrecked ships ... being in mine opinion (having small skill or judgement therein) more than would have built five of the greatest ships that I ever saw, besides mighty great boats, cables and other cordage answerable thereunto, and some such masts for bigness and length, as in mine own judgement I never saw any two could make the like.

The locals assured him that there had been twice as much wreckage at its height. In fact, however, though sizeable ships, those shattered at Streedagh, if our identifications are accurate, were by no means exceptional by Spanish standards. The *Lavia*, at 728 tons by Spanish reckoning, was the smallest ship but one in its squadron. The *San Juan* was one of the more modest of the galleons of Castille, of the currently favoured build of 500 tons or so. Their total complements were 302 and 284 men respectively. There was no other ship in the fleet that could have put the total number of victims up to the level of more than a thousand dead and three hundred survivors alleged by Cuellar. And though the locals, who assured the Lord Deputy that the beach had been littered with 1,200 or 1,300 corpses, seem to confirm Cuellar's figures, it is easy to see how, in the circumstances, the statistics might have got inflated. No atmosphere more surely breeds exaggeration than that of horrors retold. Only a dangerously *ex silentio* hypothesis—such as, that the ships had taken on crew from another vessel, or that they were accompanied by unmentioned pinnaces—could salvage Cuellar's estimates from the wreck.[12]

The two deaths which Cuellar describes in detail seem chosen for their sententious resonance. Don Diego Enríquez himself and three Portuguese nobles perished under the deck of a ship's boat, where they had ordered themselves to be battened down and sealed for the safekeeping of the worldly wealth they carried with them. The official procurator of the fleet, Martín de Aranda, whose clemency had spared Cuellar from being condemned to death and who had befriended him 'because of the great respect he had for those who are in the right', joined the narrator in an attempt to get ashore on a hatchway cover, but 'loaded with coins which he carried stitched up in his pockets', he was dragged to his death, 'crying out and calling on God while drowning'. Even in his general description of the scene of drowning and slaughter, Cuellar moralizes insistently:

I placed myself at the top of the poop of my ship, after commending myself to God and Our Lady, and from thence I gazed at the terrible spectacle. Many were drowning in the ships; others, flinging themselves into the water, sank to the bottom without returning to the surface. There were others on rafts and barrels, and gentlemen on bits of driftwood. Others cried aloud in the ships, calling on God. Captains threw their jewelled chains and crown pieces into the sea. The waves flung others away, sweeping them out of their ships. As I gazed on this appalling scene, I knew not what to do or what help to look for, as I did not know how to swim and the waves and storms were very great. And on the other side the land on shore was full of enemies, who were jumping and dancing with pleasure at our misfortune. And when any of our men reached the beach two hundred of the savages and our other enemies leapt on him and stripped him of what he wore until he was left naked in his skin.

Even before he reached land himself, Cuellar, by his own account, had narrowly escaped death four times, not counting his earlier judicial reprieve: once, through being unable to swim; next, when caught in a shower of wreckage; then when his weight plunged his hatchway-cover 'to a depth of six times my height below the surface' and at last when, with the loss of Martin de Aranda, the cover began to capsize just as a heavy spar fell, crushing and bloodying his legs. One might be forgiven for suspecting that there was a touch of Baron Munchausen about Cuellar.

The next episode was steeped in irony: he was saved by his bloody garments, for they were not worth stripping. He fell in with a pleasant youth, who was naked and dumb with shock. Two cutthroats, who approached with every appearance of evil, seemed suddenly to take pity on their prey and 'without uttering a word to us, cut a bundle of rushes and grass, covered us well and then took themselves to the shore to plunder and break open money-chests and whatever they might find, at which work more than 2,000 savages and Englishmen were garrisoned near there, were employed'. After sleeping he woke to find the young man beside him dead, 'which caused me great grief and dismay. I came to know afterwards that he was a man of consequence.' At dawn, his effort to find the Abbey of Stand led Cuellar to a scene steeped in even cruder ironies:

I found it abandoned and the church and holy images burned and utterly ruined, and twelve Spaniards were hanging there within by the hand of the Lutheran English, who went about searching for us to make an end of all who had escaped. All the monks had fled to the woods for fear of the enemy, who would have sacrificed them as well if they had caught them, as they did regularly, leaving no place of worship or hermitage standing. For they demolished them all, and made them drinking places for cows and swine.

All the themes of the rest of the narrative have now been introduced, save for the feminine interest, which Cuellar encountered later the same day,

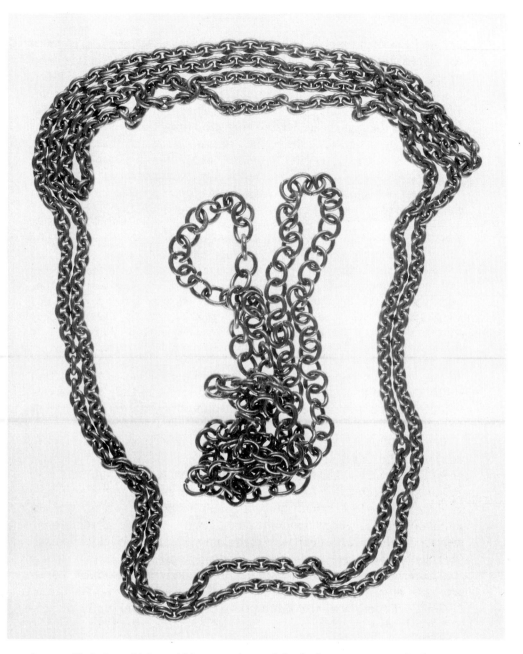

Long gold chains which could be wound round the body were a convenient means of storing wealth on one's person on board ship. At the wreck of the *Lavia*, Captain Cuellar saw stricken colleagues flinging their chains from them in an effort to reach shore. This pair are from the wreck-site of the *Girona*. The larger is 2.5 m. long and weighs nearly 2 kg.

first in the form of a melancholic old woman, who saved him from enemies who were cutting off Spaniards' heads, then a beautiful young one, who intervened to prevent his murder by her presumed lover. 'They had taken away my shirt and some relics of great value belonging to the Trinitarian Friars, which I wore inside my inner vest. They had been given to me at Lisbon. And the savage damsel took them and hung them round her neck, telling me by a sign that she wished to keep them and that she was a Christian—which she was in like manner as Muhammad.' From then on, all the native Irish Cuellar met were friendly.

For part of his journey, he had Spanish companions. His meeting with the first party of them was recounted with self-effacing humour:

I met with a lake round which there were about thirty huts, all abandoned and empty. Not knowing where to go, I sought out the best hut. On entering the door I saw it was full of sheaves of oats, which is the everyday bread these savages eat, and I gave thanks to God, but just then I saw three men emerge from one side, naked as the day their mothers gave them birth and they stood up and stared at me. They gave me a fright for I felt sure they must be devils, and they believed the same of me, girthed with my ferns and matting. As I entered they did not speak to me, because they were trembling, nor did I to them.

It was only when Cuellar uttered an exclamation against the dark that mutual recognition ensued. As if Cuellar had not already had narrow escapes enough 'they came up to me and almost finished me off with their embraces'.

Cuellar's itinerary can be briefly summarized. He was directed first to the camp of Brian O'Rourke of Breffni at Lough Glencade, near Glencar, through country where there 'were better people who, although all savages, were Christian and charitable'. O'Rourke—'although a savage, he is a very good Christian and an enemy to heretics'—was away on campaign, but seventy Spaniards had gathered at his headquarters, twenty of whom, with Cuellar in their company, set off to find a Spanish ship they had heard of 'at the coast'. Cuellar was the only one of the party not to reach her—and luckily so, he tells us, for she was wrecked soon after and all aboard her drowned or slaughtered. The narrator was now, by his own admission, lost; but a priest 'in secular clothing (for thus do the priests travel in that kingdom, so that the English shall not recognise them)' sent him to the Castle of Rossclogher of the McClancy, where he spent three months, with eight other Spanish fugitives, 'living as a real savage, just like them'. This episode is the most remarkable in the entire narrative, in which Cuellar steps out of the character he has cast for himself and becomes an urban sophisticate marooned in a sort of pastoral idyll, a Utopia of natural men, where 'the wife of the chieftain was comely in the extreme and showed me much kindness'.

One day we were sitting in the sun with some of her female friends and relatives, and they asked me about Spanish matters and other places, and in the end it came to be suggested that I should read their palms and tell their fortunes. Giving thanks to God that it had not gone worse with me than to be a gypsy among the savages, I began to study the palms of each of them and say a hundred thousand frivolous things, which pleased them so much that there was no other Spaniard better than I or higher in their favour.

Cuellar also left a famous ethnographic portrait of the Irish, rich in paradox:

The custom of these savages is to live as wild beasts among the mountains which are very rugged in that part of Ireland where we were lost. They live in huts made of peat. The men are tall, and fine in feature and member and agile as deer. They eat no more than once a day, in the evening. Their usual food is butter taken with bread made from oats. They drink sour milk, for they have nothing else. They drink no water though theirs is the best in the world. On feast days they eat meat, half-raw, without bread or salt. They wear thin trousers and short, loose coats of very coarse goats' hair. They cover themselves with cloaks and wear their hair down to their eyes. They are great walkers and indifferent to hardship. They wage perpetual war with the English who man the country for the queen. The main vocation of these folk is as robbers and to plunder each other, so that not a day goes by without some alarm among them. Most of their women are very lovely but poorly attired. They wear only a shift and a cloak, with which they cover themselves, and a linen scarf, wrapped around the head and tied in front. They are hard workers and housekeepers after their fashion. These people call themselves Christians. Mass is said among them, according to the rites of the Roman Church. Most of their churches, monasteries and hermitages have been destroyed by the hands of the English and such savages as have joined them and are as bad as they. In short, this is a kingdom where there is neither justice nor law, and everyone does what he will. The savages were well pleased with us because they knew we came to fight the heretics who were their great enemies, and had it not been for those who watched over us as if over their own lives, not a man of us would have survived. We bore them good will, therefore, although they were the first to rob us and strip naked our men who came ashore alive, from whom, and from the ruined ships of our Armada, which had so many eminent men aboard, these savages gained great riches of jewels and money.

There seem to be obvious resonances of chivalric romance and of the pastoral novel in this passage, as well as a moral message not far removed from that of Montaigne's cannibals. Yet it must not be supposed that the Spanish attitude to the Irish generally was romantic. When the Spaniards called their hosts 'savages' they meant exactly what they said. A memoirist from the *Zúñiga* used the term 'wild men of the woods', which exactly expresses the marginal humanity which was all that Spanish eyes could perceive in the Irish. The clansmen lacked almost all Thomas Aquinas's criteria of civility. Some of the practices imputed to them in Spanish accounts—the pastoral life, the raw cuisine, the inability to recognize or practise natural law, the shelter in earthern dwellings and animal pelts— are standard topoi of barbarism, with a long medieval tradition behind them. Ironically, it was as beast-men, semi-human creatures, ruled by instinct rather than reason, that the English, too, professed to classify the Irish. By that means they justified their wars against them and appropriation of their sovereignty. Much the same terms governed Spanish perceptions of the natives of the New World.

Cuellar ingratiated himself with his hosts all too well. One Sunday, after Mass, McClancy came to the Spaniards 'with hair dishevelled, down to his eyes, burning with rage' and told them he was resolved to abandon his castle and flee from an approaching English army. Cuellar protested that it was 'better to make an end honourably' and, with the other eight Spaniards, volunteered to defend the castle to the death. They fortified themselves with provisions and missiles, sacred vessels and relics, 'and the report was spread throughout the land that McClancy's castle ... would not be surrendered to the enemy because a Spanish captain, with other Spaniards who were within, guarded it'. Against a much superior force—1,800 men, by Cuellar's exaggerated account—taking advantage of the waterlogged approaches to the castle, they held out for seventeen days, until storms and snow drove the besiegers away. McClancy was so impressed that it became clear he would never let the Spaniards depart. Cuellar, with four others, therefore stole away covertly, 'one morning ten days after Christmas'. Towards the end of the month he reached the site of the wreck of the *Girona*, where he was distressed by the evidence of the deaths of so many Spanish gentlemen, but more particularly by his want of means of escape to Scotland. Hearing, however, of departures made by his countrymen from the land of the O'Cahan, he set off, once again escaping from the English by female favour at Castleroe. A boy directed him to Bishop Redmond Gallagher of Derry 'and I could not restrain tears when I approached him to kiss his hand'. The bishop had developed a reliable escape route while helping the survivors of the *Trinidad Valencera*. 'God keep him in His hands,' Cuellar exclaimed, 'and preserve him from his enemies!' He put Cuellar aboard a 'wretched boat' with other fugitives, and after a stormy passage they reached Scotland via Shetland. Cuellar did not form the same favourable impression of Scotland carried away, for example, by Juan Gómez de Medina. He felt dogged by treachery, all the way to Flanders where he supposed the Dutch were lying in wait for them off the coast by design. To avoid the Dutch, Cuellar's vessel went aground and in a macabre mockery of his first shipwreck almost exactly a year before, the narrator reached Dunkirk floating on a bit of wreckage, through a heavy sea and a high wind, 'in my shirt without any other clothing'. From the safety of the port, he saw 'the Dutch making a thousand pieces of 270 Spaniards who came in the ship which brought us to Dunkirk, not leaving three alive—for which they are now being repaid, as more than four hundred Dutchmen captured since then have been beheaded'.

As well as the ships that can be classified into groups, there were some individual vessels whose fate can be documented. One of three ships wrecked on Streedagh Strand has been identified as the Levanter *Juliana*, on the strength of a letter signed in 1596 by survivors of the *Lavia*, *Juliana*, and *Trinidad Valencera*, who were still in Ireland, serving the rebellious Ulster Catholic chiefs. The identification has been confirmed by excavations recently in progress but not yet reported in print. The laconic entry by the Secretary of the Council, Geoffrey Fenton—'in O'Flaherty's country, ships one, men 200'—together with the report of Marcos de Aramburu that until 12 September his galleon was accompanied by the '*nao Trinidad*'—seems suggestive of the Castilian support ship of that name which bore an official tally of 241 men and which does not appear to have got back to Spain. Of the hulk *Santiago*, nothing is known save the Spanish entry 'lost off Ireland' in the returned lists. A ship or ships not otherwise accounted for may perhaps be alluded to in a letter of 22 September, reporting a claim that one McLaghlen McCabb had killed eighty Spaniards from a great ship 'lost off Tralee' with his galloglass axe and in reports of Spaniards taken in the Bay of Tralee, including twenty-four 'all of Castile and Biscay, which were executed because there was no safe keeping for them'. Fenton's list includes a ship of 400 men lost in Tralee and Popham's letter of 20 September gives further confirmation. Even vaguer is the tradition of a wreck at Valentia Island, sustained by nothing more than rumour and an intriguing piece of debris in the form of a Portuguese astrolabe of appropriate date.

Of the remaining documented ships of the Armada, all but four got back to Spain without making any attempt to land. The most gallant exception was the *San Juan de Sicilia*. She was one of the half-dozen most heavily battle-damaged vessels in the fleet, emerging from the fight of 8 August battered and tattered with broken masts (above, p. 197). According to the captain of her fellow Ragusan, the *Anunciada*, to be repaired at

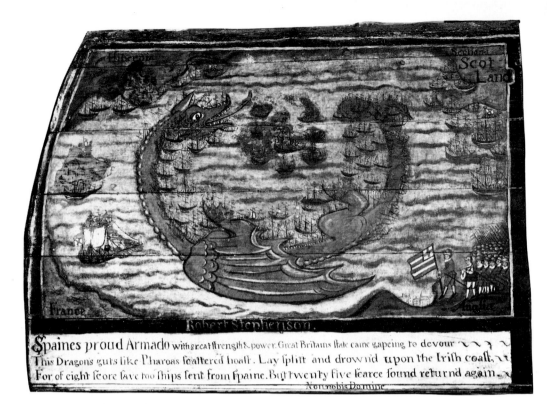

This sort of popular memorial of the Armada has survived only in small numbers, in remote places. Formerly, however, such painted panels must have been widely displayed in parish churches, fixing a particular image of the Armada conflict in English minds. In this painting on a piece of ship's board from a Lincolnshire church, the Armada's famous crescent formation is represented as a dragon. The verses are an early example of the long tradition of exaggerating the size of the fleet and the Spanish losses. Notice the prominence given by the rustic artist to his own name.

all she would first have to be taken to pieces. Yet she kept up with the main fleet until 19 August. And though thereafter laggard and alone, she continued to struggle on for Spain until 23 September, the day Medina Sidonia got home. Like the *Gran Grifón*, she was one of the leemost victims of the typhoon. And like the *Gran Grifón*, she might have been saved in consequence, for she, too, was blown not to Ireland but on to the Scottish Isles. Appearing off Islay on 23 September, she found her way to Tobermory Bay and so to the equivocal attentions of Lachlan MacLean and her end at the hands of an English agent (above, p. 15). The three remaining ships, the galleass *Zúñiga*, the hospital ship *San Pedro Mayor*, and her sister hulk *San Pedro Menor* all ended their voyages, by a strange loop of irony, back in the English Channel. Like most of the galleasses, the *Zúñiga* was failed by her rudder. After a successful halt from 16 to 23 September at an uncertain location in Ireland, where she was able to seize ample fresh provisions and water by force, she resumed her course for Spain with the first favourable wind and, despite a hazardous voyage was only fifty leagues from Corunna, by her pilots' reckoning, when, on 28 September, she was victimized by the weather's last act of malignant perversity. Forced back to the latitude of the Channel, she just managed to crawl into Le Havre on 4 October. Pedro de Igueldo, whom the king had appointed to salvage the *Santa Ana*, which had dropped out of the fleet at the very beginning of the voyage, found her 'storm beaten, with the rudder and spars broken and the ship in a sinking state ... As she was in great danger, by the favour of the governor we got her into port, not without much trouble and risk, as she was grounded at the entrance to the harbour and was within an ace of being lost.' Any other ship, indeed, in Igueldo's opinion, would have been lost. 'It would be a pity to lose so stout a piece as she is, whatever it may cost to repair her'. Indeed, despite their reputation for frailty, the galleasses in general seem to have been remarkably durable, but for their rudders. But the repairs proved a long and troublesome business especially without tools for the wrights to work. The convicts fled from their oars. Those who were French made good their escape, the rest had to be kept under guard for months with 'the greatest trouble in the world'. The men mutinied, driven by want of pay and by the frustration of an unsuccessful attempt to sail the ship home, in March 1589. The *San Pedro Menor* underwent a similar long agony, laid up at Morvieu in Brittany, from 2 October. At least, however, these two ships were in the country of a well-disposed neutral.[13]

When the *San Pedro Mayor* was driven back into the Channel, she could do no better than run aground at Hope, near Salcombe, falling straight into English hands. The late date of this disaster, which was deferred until November raises a strong presumption, unsupported, as far as I know, by any evidence, that the ship must have spent some time in Ireland on her way. But for the difficulty of salvage and the depredations of local pilferers, she might have made a good prize for Sir William Courtenay and Sir George Carey, yielding '6,000 ducats' worth of pothecary stuff'. Though Carey thought 'little good will come of the same' as it had lain in the water for a week', he was favourably impressed enough to feel well disposed towards the prisoners taken. He justified his decision to allot them $\frac{1}{2}d$. or 2d. a day each for their maintenance 'for otherwise they must needs have perished through hunger and possibly thereby have bred some infection, which might be dangerous to our country'. This sort of enlightened self-interest remained rare even on the English mainland. Carey's fellow-commissioner, Sir John Gilbert, was 'not disposed to take pain where no gain cometh' and the Spanish prisoners were only spared the fate of their compatriots in Ireland by the prudence of deferring their execution pending consideration of ransom offers. Deliverance from summary execution might well be followed by starvation in Bridewell gaol.[14]

The experience of the Armada's stragglers, looked at collectively, amply justified Medina Sidonia's warning against Ireland. Nearly two-thirds of the total number of great ships lost came to grief on that coast—at least twenty out of thirty-three. If the ships that are known to have escaped from Ireland are totted up—the four that sailed with Recalde, the six that escaped from the Shannon, the *Zúñiga*, *Nuestra Señora de Begoña* (probably), and the one or two unnamed fugitives—and contrasted with the total, including small and unidentified craft that Ireland claimed, twice as many can reliably be said to have perished there as survived. Battle damage contributed little to the losses in Ireland. Of all the Irish wrecks, only the *Trinidad Valencera*, *Rata Encoronada*, and *San Marcos* can be shown to have suffered serious battle damage, though in the cases of the *Anunciada*, the *San Juan Menor*, and the *Lavia* it might perhaps be inferred.

The intended route of the Armada according to Medina Sidonia's orders, which would have given Ireland a wide berth, was known to Robert Adams. Notice on the right the English fleet, breaking off the pursuit on 12 August and seeking refuge from the weather in East Anglian ports. The wind emblems also help to tell the story: of the southerly winds that carried the Armada up the North Sea, the north-easter that carried it round Scotland, and the south-westerly storm that wrecked the fleet, shown on the extreme left with the legend 'It was made by God.'

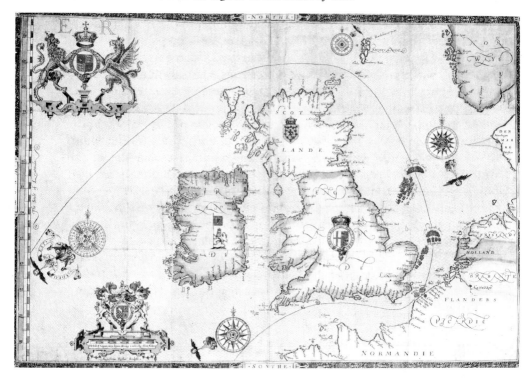

Nor does unweatherly sailing performance seem to have been a major reason for the ill-fated flight to Ireland of so many vessels: Ireland claimed six, at most, of the most unweatherly class of shipping, the hulks, doubtless because the ships were such poor sailors in a head wind that they had perforce to stand too far out to sea to make an Irish detour practicable. The Irish losses include some ships that should have been well suited to Atlantic conditions: two or three Castilians, two or three Biscayans, a Guipúzcoan, a Portuguese, as well as the *Girona*, which proved to have redoubtable sailing qualities.

Why then did so many vessels turn aside, at such great risk, and in defiance of orders? Hunger and thirst may have played some part: most of the ships that had an opportunity to do so made some attempt, at least, to take on supplies while in Ireland. Crumbling morale may have contributed to the growing atmosphere of despair in the fleet, and to make survival rank above obedience after the long, strenuous, harrowing, and frustrating voyage, of between a month and six weeks, that preceded most ships' detours. If Guipúzcoan complaints can be believed of the excesses of soldiers on their ships, frustration was vented, on the soldiers' part, in a form of violence that imperilled everyone's safety:

Seeing that they were more numerous than the seamen, they forced masters, pilots and mariners to sail the ships to the soldiers' taste and pleasure, with no consideration of what was best, taking over the rudder and tiller to achieve their own ends, and, what is more, handing out bad and harsh treatment to the seamen with threats that they would kill them, and with threatening gestures, and for this reason certain ships of the Armada have been lost, because the pilots and seafaring men had no liberty to sail them as was best.[15]

The story of the murder of the pilot of the *Rosa*, mistakenly killed for a presumed attempt at sabotage, lends some colour to this allegation. The chronology of the Irish wrecks, however, suggests that the weather was the main culprit; some ships were induced to seek shelter by the hopelessness of their efforts to make headway; others were blown there willynilly by the gales and storms of September; others were so badly stormdamaged that they had to run for cover; others driven back so far north, when nearly home, that it may have seemed wiser to pause than to press on. Armada ships only began to gather in Ireland from 12 September; until the 15th, the phenomenon was largely confined to the mouth of the Shannon. It was only in the period of the severest weather, from 17 September onwards, that ships began to be lost in large numbers: indeed, no wreck can definitely be assigned a date earlier than the 16th, when the *Trinidad Valencera* and (probably) the *San Nicolás Prodaneli* were lost.

Ten

Conclusion

THE most widespread contemporary analysis of the Armada's failure, attributing it to winds—whether of natural or providential origin—seems justified. Most of the reasons adduced by historians can be admitted in varying degrees but only in an ancillary role. It may well be true, for instance, that the Armada was foredoomed to failure from the start by errors of strategic planning and co-ordination; but, as we have seen, English strategy was equally deficient. If there was a critical omission from the Spanish plan, it was probably not, as generally alleged, the want of means of effecting a junction between the Armada and the Army of Flanders, which, though not attempted, may not have been impossible in modified circumstances, but the lack of a secure port for the Armada in northern waters. Had the plan made provisions for the acquisition of such a port, Spain's grand strategy might have been fulfilled even without the co-operation of Parma, and the subsequent disasters of exposure to crippling storms averted. Similarly, though Spanish tactics were unworkable, those of the English were ineffective. Differences in tactics made only a marginal difference to the course, and can only have made a marginal difference to the outcome, of the Armada fights. Every conceivable kind of technical deficiency has been invoked in explanation of the Armada's failure. Deficiencies in the build of Spanish ships, as we have seen, were real enough to make Spanish victory unlikely, but not bad enough to guarantee defeat; in only a few cases was weak build and bad design primarily responsible for ship losses—and even here, only to the weather and not to English guns. English superiority in the overall fighting strength of their fleet, in ordnance, quality of shot, design of gun carriages, and expertise in gunnery made some contribution, but not enough in total to disable more than three Spanish ships. The myth of generally inferior Spanish seamanship has to be dismissed altogether.

The very notion of the 'defeat' of the Armada, which has been the theme, and even the title, of so many books, has to be radically reappraised. While the two hostile fleets confronted each other, the balance of success was evenly poised between them. Spanish intentions, at the end of the day were unfulfilled, 'the outcome', in Fray Bernardo de Góngora's words, 'not as we hoped': to that extent, the English can be said to have had a victory. On the other hand the Armada escaped with a vital success to its credit: though no invasion had been launched, its feasibility had been demonstrated. Though England was unbeaten her vulnerability was assured. Everyone recognized this at the time, and showed it—the English by expecting another Armada, the Spanish by actively preparing, and eventually launching, further attempts at invasion. Only when the English fleet had been left far behind was the Armada 'defeated', and then only by the freak September weather and the reefs and rocks and cliffs of Ireland.

The experience of the Armada seems to have been much the same for those who took part in it, English and Spanish alike. They suffered together what Gonzalo in *The Tempest* called the seamen's 'common theme of woe'. The event is often supposed, however, to have marked the beginning of new and divergent courses for English and Spanish history. English success sealed the greatness of the Elizabethan age, while Spain began a descent into decline. Spanish sea power was wrecked, while that of England was launched towards ascendancy. English self-awareness was transformed by a taste of glory, while Spanish national morale began a long, cold mood of introspection and gloom. Spain's imperial career was thwarted, that of England begun. There is little to commend in these opinions, which assign the Armada far more importance than was perceived at the time or can be justified with hindsight. Wars—it has been well said—usually serve only to confirm the way things are already going, and the episode of the Armada—the indecisive encounter, the destructive aftermath— tended to confirm Spain's preponderance in Europe and hegemony in the Atlantic, which were to last at least for another two generations. Even in the immediate context of the war between England and Spain, the losses sustained by the Armada seem to have made little difference to its course or outcome. Philip II is supposed to have responded to the news of the Armada's demise with the claim that he could soon put another to sea if he wished: the incident is probably apocryphal but its implied meaning is valid. By 1592, Philip had forty galleons under construction. The Armada marked the rebirth, not the extinction, of Spanish sea power as the lost ships were replaced with better ones, and the Spanish Main refortified against attack. Throughout the decade English strikes against the Indies treasure-fleets were deflected or driven off by superior forces or even, thanks to handier Spanish shipping, outraced. In 1589, Drake abandoned his attempt to waylay the treasure and put back to Plymouth storm- damaged, wasted, and sinking; in 1590, the Spaniards stayed the treasure- ships and cheated Hawkins of his spoil. Thereafter, the English chance was gone, for Spanish strength had recovered and exceeded its former levels. In 1591 the Lord Thomas Howard's squadron abandoned its intercepting course and fled from a force of twenty galleons—except for Sir Richard Grenville, who died in the loss of the *Revenge* in an act of famous heroism or notorious folly. In 1596 an English effort to raid the Spanish Main ended in ignominy and the death of Sir Francis Drake. In the same summer, after their successful raid on Cadiz, Howard and the Earl of Essex allowed the bullion fleet to slip by safely. In 1597, Essex attempted to repair the omission, but the Spaniards outsailed him. The English were not without victories: in 1589, the English counter-Armada, led by Drake and Sir John Norris to Corunna, Lisbon, and Vigo, achieved a blow to Spanish prestige, though little in the way of material results. The Cadiz raid of 1596 was an impressive operation: a whole city seized for ransom, its harbour's shipping saved only by scuttling or flight. But the expedition was disappointed of spoil and barren of lasting results. Individual prizes fell, ever more infrequently, to English privateers in the Atlantic, but the major prizes eluded them: the Indies treasure fleet was never captured; the main Spanish battle fleets never faced, except in Grenville's futile sacrifice, much less destroyed.

Spain's reply by sea in the same period was not much more effective, though in the second half of the decade, as we have seen, the Spaniards raided Cornwall with impunity and Spinola carried counter-piracy into the Channel. The menace, however, of Spanish sea power was stronger after the Armada than before. The events of 1588, though they had ended badly for Spain for reasons little connected with English power, had demonstrated the weakness of English defences. The best measure of the Armada's success was that Philip II was willing to try it again. The Armada of 1596 was driven back by storms; that of 1597 sailed with a rational plan, based on the securing a safe haven at Falmouth; the fleet was overwhelmingly powerful—outnumbering the English in fighting ships of all classes, attaining an 'invincibility' which that of 1588 had possessed in nothing but name—and was descending on an undefended coast, when the weather again intervened to the English advantage and dispersed it in the very mouth of the Channel.

Apart, however, from the formidable exchange of blows in 1596–7, there was an air of phoney war about Anglo-Spanish hostility after the Armada. No mortal thrust was again attempted by either side. The Armada did nothing to increase either side's belligerency. The main theatre of conflict, the main focus of interests hard or impossible to reconcile, was still the Low Countries; here Elizabeth continued to pursue strictly limited objectives and fought, in large part, a war of subsidies, with vicarious forces, French and Dutch. The wider, sea-borne war reverted more or less to the level of the early 1580s, fought selectively and reluctantly. It remained unpopular because of the damage to commerce. *The Marchants Avizo* of 1589, reprinted in 1590 and 1591, was an English merchants' manual, chiefly concerned with the Spanish trade, and an eloquent testimony to determination of English traders to resume normal commercial relations. Pressure from Spain was almost as strong to bring the wasteful conflict to an end. The Spanish merchants trading from Cadiz lost 12 million ducats' worth of goods in the conflict of 1596. In 1594, rich men aboard the East Indiaman *Cinco Llagas* lost their lives when the captain refused to heed their pleas for surrender to English pirates. These men and their successors set little value on revenge. The war grew more unpopular as it grew more expensive, particularly in England towards the end of Elizabeth's reign when it became apparent that it could not be continued without extraordinary taxation. The relief with which peace was greeted can be sensed vividly by anyone who climbs the stairs of the Bodleian Library, as I have done most days of my adult life, and sees the great frieze in which James I, the peacemaker, is depicted in triumph, crowned by the blessings of peace. Shakespeare, who is often cited as the voice of the new chauvinism excited by English success, in Sonnet 107 celebrated the survival of 'the mortal moon', and the coming of 'olives of endless age'.

This is not to say that pride in achievement was not stirred in England by deliverance from the Armada, or by the propaganda which it occasioned. And that pride may have contributed something to the development of national self-awareness and self-consciousness. In Bohemia, the magus John Dee and his medium, hearing the news by ordinary means, felt an urge to return to 'Your British Earthly Paradise and Monarchy Incomparable'. According to the *Letter to Mendoza*, the level of popular rejoicing at home was impressive: captured Spanish banners and streamers were 'brought to Pauls Church-yard, and there showed openly in the sermon time, to the great rejoicing of all the people. And afterwards they were carried to the crosse in Cheap, and afterwards to London Bridge, whereby the former rage of the people was greatly assuaged.' Churches, Burghley claimed, 'filled daily to hear sermons wherein is remembered the great goodness of God towards England'. Such officially sponsored junkets are, of course, no genuine guide to popular feeling, any more than the Armada portraits, jewels, and medals that so effusively extolled the queen and reviled the enemy. There have survived, however, some suggestive examples of enthusiasm at lower social and educational levels. The art of the balladeers produced *A Joyful New Ballad*

Relief at the coming of peace is expressed in this image over the gate of the Bodleian Library. King James is being extolled by Victory and adored by Piety, while he distributes the fruits of his learning and munificence. The instruments of war—arrow, armour, sword, axe, and gun—are trampled beneath his feet, while the baldocchino is engraved 'Blessed are the Peacemakers.'

Declaring the Happy Obtaining of the Great Galleazzo, wherein Don Pedro de Valdes was Chief, and *The Queen's Visiting of the Camp at Tilbury, with her Entertainment there* was sung to the tune of 'Wilson's Wild'. In a parish church, in the Lincolnshire marshland—not far, as the crow flies, from the battlefields of the Netherlands—hang some planks of ship's timber painted by one Robert Stephenson with the Armada's familiar crescent formation twisted into the shape of a loop-tailed dragon, whose 'guts', according to the inscription, 'like Pharaoh's scattered host,/ Lay split and drowned upon the Irish coast'. A similar, but less naïve, panel in St Faith's Church, Gaywood, near King's Lynn, gives pride of place to the figure of the queen, shown among her troops at Tilbury, and at prayer, in a sanctum at the top of the diptych, in gratitude for the divine grace wielded against the Armada. The Victoria and Albert Museum has a cast-iron fireback made in England in 1588 with nautical embellishments, perhaps in celebration of the same event. These modest works of verse and art are freak survivals. There must have been a great deal more such ephemera produced at the time. Armada drinking-glasses and even Armada playing-cards have survived from well into the seventeenth century and may form part of a tradition of celebratory bric-à-brac begun when the impact of the event was fresh.

The boost to English morale, such as it was, is impossible to calibrate. It is hard even to detect any corresponding depression, traceable to the influence of the Armada, in morale in Spain. It is true that the 1590s were, on the whole, a decade of frustration and therefore introspection as Spanish poets and projectors anatomized the monarchy's ills and sought explanations and remedies for apparently stymied success. The Armada, however, seems to have played little or no part in the creation of this mood. On the contrary, the problems of Spain in the 1590s arose not from depressed morale but from excited expectations. This was a decade in which the conquest of France was attempted and that of China projected; in which the defences of the monarchy were modernized from Manila to Milan; in which three more Armadas were assembled, and two launched, and in which the crown's efforts were supported by almost the entire monarchy, with the exception of the Netherlandish rebels and, briefly, some pockets of disaffection in Aragon. Failure in these circumstances does not appear as the consequence of undermined self-confidence, but rather as 'the inevitable effects of immoderate greatness' and of resources overreached. By analogy with the 'generation of 1898'— a well-established concept of Spanish historiography—said to have been made aware of Spain's eclipse as a great power by defeat in war against the United States, a distinguished Spanish scholar has recently invented the concept of a 'generation of 1588', bred up to a sense of failure by childhood memories of the Armada. In fact, outside strictly historical works, the Armada is rarely recalled in Spanish literature of the early seventeenth century. It might, perhaps, be appropriate to speak of a 'generation of 1598', by which time the failure in France, the stagnation of Spanish progress in the Netherlands, the weakness of Spanish power in the Pacific, the dispersal of the second and third Armadas, and the lessons of Essex's outrage at Cadiz had all become apparent. In that year, the monarchy's retreat from its summit of ambition was signalled by the peace of Vervins with France, recognizing the status quo, and by the political testament of the dying Philip II, which urged moderation and compromise on his heir. In the immediate aftermath of the Armada, none of this had been foreshadowed. On the contrary, as we have seen, the prevailing mood was defiant. Within a few days of the arrival of the remnant of the fleet, the *Junta de Noche* was at work on preparations for the next Armada.

Whether in the longer term the outcome of the Armada made any difference to the imperial fortunes of England and Spain, or to the global balance of sea power, must also be doubted. 'Heaven's command', was not uttered in 1588—or, if it was, it was not heeded. As we have seen, the English fleet, despite enormous advantages 'of time and place', was unable to defeat the Armada or to demonstrate convincing command of England's home waters. The lesson contemporaries drew from the experience of the Armada, made palpable in the form of the successor-Armadas, was that England was vulnerable to invasion, even by a distant enemy with no safe harbour at hand. English naval hegemony, which was so marked a feature of world history in the eighteenth and nineteenth centuries, was still at least a hundred years off. Spanish naval strategy was still without a solution to what was perhaps its fundamental problem—its lack of a base in northern waters; but a solution continued to be sought until the end of the Thirty Years' War. In the aftermath of the Armada, no English empire was created, except in Ireland; nor did the Spanish empire lose, in Philip II's words 'a single oratory where the name of Christ was praised'. The New World empire England built up in the seventeenth century proved less durable than that of Spain. As for the character of naval warfare, as we have seen, the view that the Armada encounters were tactically innovative, or marked a new departure in the history of naval strategy, is hard to justify in the context of a gradual process of evolution towards the supremacy of gunnery and of the strategy of long-range blockade. The most that can be said is that the Armada was a step in that process. It is contrary to common sense to ascribe transcendent

The 2ᵈ Fight betweene ẏ Engliſh and Spaniſh Fleetes being the 23 of Iune 1588. wherein only Cock an Engliſhman being wᵗʰ his litle Veſſell in ẏ midſt of ẏ Enemies died valiently, but ẏ Spaniards much worſted.

This scene on an Armada playing-card evokes the essence of the most precious English myth of the event—that of a heroic English struggle against the odds, exemplified here by the fabulous Cock 'with his little vessel in the midst of the enemies'.

influence, other than of a cautionary kind, to a conflict so indecisive, in which the strategies and tactics tried on both sides seemed to have failed.

The enduring influence of the Armada has been felt in the realm of myths not made at the time but slowly accumulated from the accretions of a long historical and literary tradition: the myths of a great English victory, of English superiority over Spain; of the outcome of the Armada as a symbol of an age of English national greatness in the reign of Elizabeth I; of the Armada fight as part of a war of religion. These myths are the last stragglers of the Armada, and have still to come into port. Future generations of embarrassed schoolboys and critical readers will, no doubt, continue to squirm as long as the poems of Newbolt are read, or *Merrie England* performed, or pageants enacted, like that of E. F. Benson's *Lucia*, at old-fashioned garden fêtes. The myths—provided they are not believed—have an innocent charm of their own, and the art that expresses them is often noble and inspiring and properly proud. It would be churlish to hope, cruel to seek, for their demise. On the other hand, it can do them no harm to be seen—as they will increasingly be seen—in the context of the reality of the Armada and, in particular, of the shared experience of the muddle and misery of war that was common to the antagonists on both sides. That was the contemporary—and, therefore, the proper historical—significance of the passage of arms between English and Spanish sea-borne forces in 1588. As a result, the Armada will in future seem less important than formerly, but not, I think, less interesting.

Notes

Abbreviations used in the Notes

Boynton	L. Boynton, *The Elizabethan Militia* (London, 1967).
Coco Calderón	J. Paz (ed.), 'Relación de la "Invencible" por el contador Pedro Coco de Calderón', *Revista de archivos, bibliotecas y museos*, 3rd s., 1 (1897).
Fernández Duro	C. Fernández Duro, *La armada invencible*, 2 vols. (Madrid, 1884–5).
Herrera Oria *et al.*	*La armada invencible: Documentos procedentes del Archivo General de Simancas*, ed. E. Herrera Oria, M. Bordouan, and A. de la Plaza (Valladolid, 1929).
Hume	*Calendar of Letters and State Papers Relating to English Affairs, Preserved Principally in the Archives of Simancas*, ed. M. A. S. Hume, 4 vols. (London, 1892–9).
Laughton	*State Papers Relating to the Defeat of the Spanish Armada*, ed. J. K. Laughton, 2 vols. (London, 1894–5).
Lomas	*Calendar of State Papers, Foreign*, ed. S. C. Lomas (London, 1927).
Lyell	J. P. R. Lyell, 'Commentary on Certain Aspects of the Armada', Bodleian Library MS.
Martin	C. Martin, *Full Fathom Five* (London, 1975).
Maura	G. Maura y Gamazo, *El designio de Felipe II* (Madrid, 1957).
Naish	G. B. Naish (ed.), 'Documents Illustrating the History of the Spanish Armada', *The Naval Miscellany*, 4, ed. C. Lloyd (London, 1952).
Sténuit	R. Sténuit, *Treasures of the Armada* (London, 1974).
Van der Essen	L. van der Essen, *Alexandre Farnèse*, 5 vols. (Brussels, 1933–7).

Preface

1. Herrera Oria, *et al.*, p. 230.
2. See e.g. Fernández Duro, esp. ii. 331; M. Fernández Álvarez, *La sociedad española en el siglo de oro* (Madrid, 1983), 707.
3. For the 1888 celebrations see W. H. K. Wright, *Catalogue of the Exhibition of Armada and Elizabethan Relics* (London, 1888).
4. So far the only piece produced on the basis of the neglected Simancas material has been I. A. A. Thompson, 'The Spanish Armada Guns', *The Mariner's Mirror*,

61 (1975), 355–71. On the Dutch archives see J. B. van Overeem, 'Justinus van Nassau en de Armada', *Marineblad*, 53 (1938), 821–31, and G. Mattingly, *The Defeat of the Spanish Armada* (London, 1959), 439. I am conscious that my own ignorance of these sources may make me unduly dismissive of the importance of the Dutch (pp. 114–16). The Medina Sidonia archives were used, to little apparent effect, by P. Pierson, 'A Commander for the Armada', *The Mariner's Mirror*, 55 (1969), 383–400. Pending their re-emergence, many extracts from their contents can be found in Maura. All but one of the excavated wrecks are reported in Sténuit, Martin and S. Wignall, *In Search of Spanish Treasure* (Newton Abbot, 1982).

Chapter One

1. Cervantes, *Viaje del Parnaso*, ch. 1 (Madrid, 1614), f. 1. This and the next three paragraphs are based on documents in L. Astrana Marín, *Vida ejemplar y heróica de Miguel de Cervantes Saavedra*, 7 vols. (Madrid, 1948–57), iv. 255–63. I am indebted to W. Byron, *Cervantes* (London, 1979), 322–38.
2. Hume, iv. 230, 323–4; Lyell, p. 130; Fernández Duro, i. 385, ii. 134–7.
3. J. S. Corbett, *Papers Relating to the Navy during the Spanish War, 1585–87* (London, 1895), 111; R. Quatrefages, 'La proveeduría des armades, 1535–41', *Mélanges de la Casa de Velázquez*, 14 (1978), 217. On the plans of Santa Cruz in general, see E. Herrera Oria, *Felipe II y el Marqués de Santa Cruz en la empresa de Inglaterra según los documentos . . . de Simancas* (Madrid, 1946).
4. Fernández Duro, i. 249; see below, p. 130.
5. M. and C. Silveira, 'A alimentaçao na Armada invencivel', *Revista de história* (São Paolo), 36 (1968), 306–7; A. Crinò, 'La partecipazione di un galeone del Granduca di Toscana Ferdinando I all'impresa d'Inghilterra di Filippo II di Spagna nel 1588', *Archivio storico italiano*, 142 (1984), 597; Herrera Oria *et al.*, pp. 384–435; Archivo General de Simancas, Guerra Antigua, leg. 221, f. 152.
6. Lomas, xxi, part IV, p. 208; Hume, p. 375; below, p. 222.
7. N. Sánchez-Albornoz, 'Gastos y alimentación de un ejército en el siglo XVI', *Cuadernos de historia de España*, 14 (1950), 153; *Calendar of State Papers, Venetian*, ed. H. F. Brown (London, 1894), viii. 334–6.
8. M. Ulloa, *La hacienda real de Castilla en el reinado de Felipe II* (Madrid, 1977), 807; Hume, iv. 224.
9. Laughton, i. 191.
10. Crinò, 'La partecipazione di un galeone del Granduca', pp. 589–90, 600–2.
11. J. de Courcy Ireland, 'Ragusa and the Spanish Armada', *The Mariner's Mirror*, 64 (1978), 256–61.
12. F. F. Olesa Muñido, *La organización naval de los estados mediterráneos y en especial de España durante los siglos xvi y xvii*, 2 vols. (Madrid, 1968), i. 523; Fernández Duro, i. 424, ii. 201–3; Naish, p. 10; Lyell, pp. 320–1.
13. Laughton, i. 134, 252–3; Boynton, pp. 125, 154–5; M. Lowther, *Survey of the Coast of Sussex* (Lewes, 1870).
14. Laughton, i. 143–4, 146–7, 153–5; Lyell, p. 95 n. 2; *Acts of the Privy Council*, ed. J. R. Dasent (London, 1897), xvi. 113, 123–4, 159–62, 165.

15. *Acts of the Privy Council*, xvi. 152, 155–9.
16. N. Alonso Cortés, *Miscellanea valisoletana*, 3 (Valladolid, 1921), 23–33; J. M. Lope Toledo, 'Logroño en el desastre de la armada invencible', *Berceo*, 17 (1962), 231–43.
17. Boynton, pp. 42, 44, 48, 148; Laughton, i. 159.
18. Laughton, i. 54, ii. 187; Hume, iv. 466–8; Fernández Duro, ii. 475; Lyell, p. 123 n.; C. G. Cruickshank, *Elizabeth's Army* (Oxford, 1966), 30.
19. Hume, iv. 466–8.

Chapter Two

1. Herrera Oria *et al.*, p. 211; C. G. Cruickshank, *Elizabeth's Army* (Oxford, 1966), 29.
2. *Vida, nacimiento, padres y crianza del capitán Alonso de Contreras*, ed. F. Reigos (Madrid, 1967); R. Quatrefages, *Los tercios españoles* (Madrid, 1979), 319.
3. 'Carta relativa a la armada invencible', *Revista de archivos, bibliotecas y museos*, 1st s., 2 (1872), 368–9; Fernández Duro, i. 498.
4. Fernández Duro, i. 74, 473, 497–8, 502; Laughton, i. 304–5. On the Spanish chivalric tradition at sea see F. Fernández-Armesto, *Before Columbus* (London, 1987), 3, 74.
5. Lyell, p. 172; Lomas, xxi, part i, p. 12.
6. *Cartas de Eugenio de Sálazar* (Madrid, 1866), 42; Hume, iv. 294–5.
7. Herrera Oria *et al.*, pp. 188–9; Hume, iv. 294–5.
8. Herrera Oria *et al.*, p. 88; Naish, p. 23; Hume, iv. 475–6.
9. *Colección de documentos inéditos para la historia de España*, 112 vols. (Madrid, 1842–95), lxxxi (1883), 200; Fernández Duro, ii. 314–15; Maura, pp. 250, 274; Herrera Oria, *et al.*, p. 153; Hume, iv. 266.
10. Maura, p. 275; *Calendar of State Papers, Venetian*, ed. H. F. Brown (London, 1894), viii. 351.
11. *Acts of the Privy Council*, ed. J. R. Dasent, xv (London, 1897), 414; C. Weiner, 'The Beleaguered Isle', *Past and Present*, 51 (1971), 27–62; Laughton, i. 58, 61; D. Archdeacon, *A True Discourse of the Armye which the King of Spaine caused to be Assembled* (1588), 5.
12. Lyell, p. 172; *The Copie of a Letter . . . found in the Chamber of Richard Leigh* (London, 1588); Naish, p. 63; Boynton, pp. 100, 149; on recusants generally, W. R. Trimble, *The Catholic Laity in Elizabethan England* (Cambridge, Mass., 1964); A. J. Gerson, 'The English Recusants and the Spanish Armada', *American Historical Review*, 22 (1917), 589–94.
13. A. L. Rowse, *Eminent Elizabethans* (London, 1974), 130; H. Kamen and J. Peres, *La imagen internacional de la España de Felipe II* (Valladolid, 1980), 55–6.
14. L. de Góngora, *Obras poéticas*, 3 vols. (New York, 1921), i. 109; Fernández Duro, i. 238.
15. Naish, p. 60; J. Bruce, *Report of the Arrangements which Were Made for the Internal Defence of These Kingdoms* (London, 1798), p. ccjviii; T. Deloney, *A New Ballet of the Strange and Most Cruel Whips* (London, 1588); Lyell, pp. 190–1.
16. Hume, iv. 294–5; Lomas, p. 341; Laughton, ii. 342; Fernández Duro, ii. 273.

Chapter Three

1. G. Birmingham, *Spanish Gold* (London, 1973), 64; Sténuit, pp. 189–229.
2. Fernández Duro, i. 473, ii. 474
3. Maura, pp. 275–6; *Cartas de Eugenio de Sálazar* (Madrid, 1866), 46–52.
4. Martin, p. 116; Sténuit, p. 190; Fernández Duro, ii. 291.
5. *Autobiografías de soldados*, ed. J. Cossió (Madrid, 1956), 7, 27.
6. F. F. Olesa Muñido, *La organización naval de los estados mediterráneos y en especial de España durante los siglos XVI y XVII*, 2 vols. (Madrid, 1968), ii. 721–2; *Colección de documentos inéditos para la historia de España*, 112 vols. (Madrid, 1842–95), lxxxi (1883), 194.
7. L. Van der Essen, 'Documents concernant le Vicaire Général Francesco de Umara', *Analectes pour servir à l'histoire ecclésiastique de la Belgique*, 3rd s., 7 (1911), 263–81.
8. R. Quatrefages, *Los tercios españoles* (Madrid, 1979), 535–7.
9. C. G. Cruickshank, *Elizabeth's Army* (Oxford, 1966), 145; Fernández Duro, i. 426–8.
10. Hume, iv. 266; Quatrefages, *Los tercios españoles*, pp. 271, 308.
11. *Historical Manuscripts Commission: Calendar of the MSS of the Most Honourable the Marquess of Salisbury preserved at Hatfield House*, ii (1888), 222; *Cartas de Eugenio de Sálazar* (Madrid, 1866), 143.
12. Cruikshank, *Elizabeth's Army*, p. 183; Hume, iv. 295; N. Sánchez-Albornoz, 'Gastos y alimentación de un ejercito en el siglo XVI', *Cuadernos de historia de España*, 14 (1950), 167; B. Vincent, 'Consommation alimentaire en Andalousie orientale 1581–2' *Annales*, 30/23 (1975), 445–53; Fernández Duro, i. 425; Laughton, ii. 159–60; J. Schoonjams, ' "Castra Dei": l'organisation religieuse des armées', *Miscellanea historica in honorem Leonis Van der Essen*, 2 vols. (Brussels, 1947), i. 523–40.
13. Hume, iv. 269–70; *Calendar of State Papers, Domestic* (1581–90), ed. R. Lemon (1865), ccix. 16; M. and C. Silveira, 'A alimentaçao na Armada invencivel', *Revista de história* (São Paolo), 36 (1968), 304, 309.
14. Fernández Duro, i. 353.
15. M. and C. Silveira, 'A alimentaçao na Armada invencivel', pp. 311–12.
16. Herrera Oria *et al.*, p. 230; Laughton, ii. 222, 226; Fernández Duro, ii. 474.
17. *Cartas de Sálazar*, p. 45; Martin, p. 123; Fernández Duro, i. 286; Quatrefages, *Los tercios españoles*, p. 61; Vincent, 'Consommation alimentaire en Andalousie orientale 1581–2', pp. 448–9; Laughton, i. 190, 334, ii. 303–4.
18. Laughton, ii. 95; Fernández Duro, ii. 470.

Chapter Four

1. *The Deposition of Don Diego Pimentelli* (London, 1588); Lyell, pp. 191–2.
2. C. Fernández Duro, i. 243; Lomas, xxi, part I, p. 365; Van der Essen, v. 165; Hume, iv. 28, 385, 452.
3. *Calendar of State Papers, Venetian*, ed. H. F. Brown (London, 1894), viii. 182, 189, 338.

4. Van der Essen, v. 175–6; Maura, p. 167; Hume, iv. 233–7; Lyell, pp. 191–2.
5. Fernández Duro, i. 241–4; V. Fernández de Asis, *Epistolario de Felipe II sobre asuntos de mar* (Madrid, 1943), 779.
6. Fernández Duro, i. 23; Van der Essen, v. 16; *Correspondence of Robert Dudley, Earl of Leicester*, ed. J. Bruce (London, 1944), p. 312.
7. Fernández Duro, i. 244–7, 320.
8. Van der Essen, v. 163–4, 166, 173.
9. Lyell, pp. 17–21; Van der Essen, v. 169, 178–9; Naish, p. 8.
10. Hume, iv. 136–7; Van der Essen, v. 188; E. Herrera Oria *et al.*, pp. 56–7, 59, 79.
11. Lomas, xxi, part IV, pp. 371, 552; Laughton, i. 301, ii. 46; Van der Essen, v. 160; Herrera Oria *et al.*, pp. 19, 131.
12. Van der Essen, v. 200–2; Hume, v. 211, 233–7; Laughton, i. 46–7.
13. See Boynton, p. 141.
14. Ibid. 132, 135; *Calendar of State Papers, Venetian*, ix. 237; *Acts of the Privy Council*, ed. J. R. Dasent (London, 1897), xvi, 302.
15. J. Bruce, *Report on the Arrangements which were made for the Internal Defence of these Kingdoms* (London, 1798), p. xlviii; Boynton, pp. 145–7.
16. Laughton, i. 123–6; Maura, p. 245.
17. Laughton, i. 60, 192–3, 200, 237–43.
18. Ibid. ii. 202–5.
19. Ibid. ii. 96.
20. P. Pierson, 'A Commander for the Armada', *The Mariner's Mirror*, 55 (1969), 383–400.

Chapter Five

1. Naish, p. 8.
2. Herrera Oria *et al.*, pp. 59, 75, 112, 123, 129, 134–6.
3. Maura, p. 121.
4. Fernández Duro, i. 368.
5. *Colección de documentos inéditos para la historia de España*, 112 vols. (Madrid, 1842–95), lxxxi (1883), 233; Herrera Oria *et al.*, p. 152; Naish, p. 12.
6. Naish, p. 18; M. de Foronda y Gómez, *Estudios del reinado de Felipe II* (Madrid, 1954), 207; Hume, iv. 249.
7. Herrera Oria *et al.*, pp. 24, 202, 325, 334.
8. Ibid. 186, 220, 248, 331, 379; Hume, iv. 321–4.
9. Naish, pp. 38, 53; Laughton, i. 202.
10. Laughton, i. 191–2, 204–6, 292, 299–300; Boynton, p. 141; British Library Lansdowne MS 1225, fos. 23–40; Lyell, p. 184.
11. *Calendar of State Papers, Domestic* (1581–90), ed. R. Lemon (1865), ccxii, no. 52; Lyell, pp. 83, 90–1; Lomas, xxi, part IV, p. 511.
12. Fernández Duro, ii. 134–7; Herrera Oria *et al.*, pp. 192, 196, 211.
13. Van der Essen, v. 198–9; Laughton, i. 341, ii. 49; Herrera Oria *et al.*, p. 5; *Acts of the Privy Council*, ed. J. R. Dasent (London, 1897), xvi. 210.
14. Naish, p. 23; Hume, iv. 202–3; Laughton, ii. 23.
15. Herrera Oria *et al.*, pp. 193, 373; Hume, iv. 249, 357, 370; Van der Essen, v. 185, 188; Lyell, p. 184; Laughton, i. 332, ii. 28; Naish, p. 20.

16. Hume, iv. 310.

17. Herrera Oria *et al.*, pp. 131–2; Lyell, pp. 29–37; Naish, pp. 15–16; Hume, iv. 245–9.

18. Laughton, ii. 28, 133; Van der Essen, v. 204; Fernández Duro, ii. 185.

19. Van der Essen, v. 187; Hume, iv. 135–7, 250.

20. Lyell, pp. 38–40; C. V. Malfatti, *Cuatro documentos italianos en materia de la expedición de la armada invencible* (Madrid, 1972), p. 12; Naish, p. 11; Herrera Oria *et al.*, pp. 196–7; Hume, iv. 365.

21. Herrera Oria *et al.*, p. 12; Hume, v. 334–5, 347; Laughton, ii. 133.

22. Hume, iv. 315–16.

23. Fernández Duro, i. 410, 417, 489; Hume, iv. 299.

24. Maura, pp. 172, 243; Hume, iv. 207–8; Laughton, i. 134; Fernández Duro, iii. 92; R. Gray, 'Spínola's Galleons in the Narrow Seas', *The Mariner's Mirror*, 44 (1978), 71–83.

25. Hume, iv. 357–8, 362–4, 366.

26. Hume, iv. 370.

27. Van der Essen, v. 204; Lyell, pp. 140–1.

28. A. Vázquez, *Los Sucesos de Flandes* (*Colección de documentos inéditos para la historia de España*, lxxii–lxxiii), 2 vols, ii. 348–9.

29. Hume, iv. 370; Laughton, ii. 99.

30. *Correspondance du cardinal de Granvelle*, ed. E. Poullet and C. Piot, 12 vols. (Brussels, 1877–89), xii. 487; Conyers Read, *Mr Secretary Walsingham*, 3 vols. (London, 1925), iii. 265–6; Laughton, ii. 20, 51; Van der Essen, v. 161–2, 182, 188, 197; Hume, iv. 261.

Chapter Six

1. G. Birmingham, *Spanish Gold* (London, 1973), 30; Lyell, p. 140; Laughton, ii. 259.

2. Maura, p. 176.

3. Hume, ii. 480, iv. 247; Lyell, pp. 129–37; 141; Herrera Oria *et al.*, p. 132; G. Mattingly, *The Defeat of the Spanish Armada* (London, 1959), 233.

4. Medina Sidonia's narrative for 22 July–7 August is printed in *Colección de documentos inéditos para la historia de Espana*, 112 vols. (Madrid, 1842–95), xiv (1849), 449–61, and completed ibid. xliii (1863), 417–25 for 8–20 August. Reproduced or translated with varying degrees of fidelity in several collections, it is arranged chronologically, so that citations are easily verified and will not be given individually in these notes. Hume, iv. 373; Herrera Oria *et al.*, p. 268; Laughton, i. 355.

5. D. García del Palacio, *Instrucción naútica para navegar* (Mexico, 1587), 120–7.

6. Hume, iv. 360; Lyell, p. 324; Laughton, ii. 102; D. W. Waters, 'The Elizabethan Navy and the Armada Campaign', *The Mariner's Mirror*, 35 (1949), 122.

7. Hume, iv. 293; Laughton, i. 288, ii. 59, 64, 207.

8. Laughton, i. 340–1, 345, 207; *Acts of the Privy Council*, ed. J. R. Dasent (London, 1897), xvi. 209.

9. J. S. Corbett, *Fighting Instructions 1530–1816* (London, 1905), 33–5, 41–3; Waters, 'The Elizabethan Navy', p. 122.

10. Coco Calderón, pp. 32, 34; Lyell, p. 371.

11. *Calendar of State Papers, Domestic* (1591–4), ed. M. A. E. Green (1867), 289; Laughton, i. 85–6, ii. 78; Lyell, p. 141.

12. Hume, iv. 327–8, 349; Martin, p. 133; Fernández Duro, i. 452; Waters, 'The Elizabethan Navy', p. 199.

13. Hume, iv. 364–5; C. V. Malfatti, *Cuatro documentos italianos en materia de la expedición de la armada invencible* (Madrid, 1972), p. 42; *Colección de documentos*, ii. 171.

14. Sir W. Raleigh, *Works*, viii. 237; on Hawkins and shipbuilding generally, see M. Oppenheim, *History of Naval Administration 1509–1660* (London, 1896), and J. A. Williamson, *The Age of Drake* (London, 1965), 253–72.

15. Fernández Duro, i. 439, 447; Martin, pp. 129–30; Thomas Nash, *Pierce Pennilesse His Supplication to the Devil* (London, 1592); Lyell, p. 107.

16. Laughton, i. 12, 289, 323–4, ii. 60; Malfatti, *Cuatro documentos italianos*, p. 42.

17. Laughton, ii. 50, 54, 134, 224, 259.

18. Ibid. ii. 40; Lyell, p. 116 n.; Calendar of the Pipe Office Accounts, 1588, no. 2225.

19. Laughton, ii. 102, 241–54; Naish, p. 72.

20. M. Lewis, *The Armada Guns* (London, 1961); I. A. A. Thompson, 'The Spanish Armada Guns', *The Mariner's Mirror*, 61 (1975), 365–71.

21. See also N. A. M. Rodger, 'Elizabethan Naval Gunnery', *The Mariner's Mirror*, 61 (1975), 353–4; with this addition, the figures for range and calibre are based on Lewis.

22. Martin, pp. 207, 212; Laughton, ii. 21; *Colección de documentos*, lxxxi (1883), 190; Herrera Oria, *et al.*, p. 136.

23. Lomas, pp. 58, 93; Lewis, *The Armada Guns*, pp. 106, 137 (but cf. Thompson, 'The Spanish Armada Guns', p. 362); Martin, p. 175.

24. S. Wignall, *In Search of Spanish Treasure* (Newton Abbot, 1982), 88; E. W. Bovill, 'Queen Elizabeth's Gunpowder', *The Mariner's Mirror*, 33 (1947), 179–86.

25. Hume, iv. 360; 401, 442; Lyell, p. 326; Fernández Duro, ii. 376–7; Laughton, i. 13.

26. García del Palacio, *Instrucción naútica para navegar*, pp. 121–2; Hume, iv. 442; *Colección de documentos*, lxxxi (1883), 190; Martin, pp. 215–16.

27. Fernández Duro, ii. 376–8; Hume, iv. 468; Lyell, p. 140.

28. *Acts of the Privy Council*, ed. Dasent, xvi. 272; Lyell, pp. 348–9; Laughton, i. 11, 13.

29. Lyell, pp. 145–6; Herrera Oria *et al.*, pp. 351–5.

30. Hume, iv. 321–4; *Colección de documentos*, lxxxi (1883), 231.

31. Herrera Oria *et al.*, p. 331; Laughton, ii. 106.

32. Coco Calderón, p. 33.

33. Naish, p. 16.

34. Hume, iv. 316–17, 329–30, 392; Laughton, ii. 23; Naish, p. 45.

35. Laughton, i. 190, ii. 373–4.

Chapter Seven

1. Laughton, i. 15; 321–2.

2. Herrera Oria *et al.*, pp. 182, 198.

3. Laughton, ii. 8; D. W. Waters 'The Elizabethan Navy and the Armada Campaign', *The Mariner's Mirror*, 35 (1949), 122; Calendar of the Pipe Office Accounts, 1588, no. 2225.
4. Coco Calderón, p. 35; Herrera Oria *et al.*, p. 257; Lyell, p. 329.
5. Fernández Duro, ii. 269, 283.
6. Laughton, ii. 10.
7. Ibid. ii. 64; Lyell, pp. 325–6; Hume, iv. 396; Coco Calderón, p. 36; Herrera Oria *et al.*, p. 301.
8. Coco Calderón, pp. 36–7; J. de Courcy Ireland, 'Ragusa and the Spanish Armada', *The Mariner's Mirror*, 64 (1978), 259–60; Laughton, ii. 2, 10–11, 80.
9. Fernández Duro, ii. 406.
10. Coco Calderón, p. 38; Hume, iv. 376–7, 459–62; Lyell, pp. 140, 325; M. Lewis, *The Armada Guns* (London, 1961), 192–3.
11. Maura, p. 263.

Chapter Eight

1. Hume, iv. 377.
2. Coco Calderón, p. 38; Herrera Oria *et al.*, p. 325.
3. Laughton, ii. 38.
4. Ibid. ii. 12.
5. Fernández Duro, ii. 395.
6. Laughton, ii. 94.
7. Coco Calderón, p. 38; Fernández Duro, ii. 337–70, 389–98.
8. Coco Calderón, p. 38; Fernández Duro, ii. 278; Maura, p. 265.
9. Fernández Duro, ii. 408–9; Lyell, p. 327.
10. Herrera Oria *et al.*, p. 352; Hume, iv. 456–7.
11. Laughton, ii. 41–2.
12. Ibid. ii. 98; Coco Calderón, p. 39.
13. Fernández Duro, ii. 278; Lyell, pp. 143, 328.
14. Fernández Duro, ii. 253, 289; Laughton, ii. 142–7, 173.
15. Fernández Duro, ii. 253, 278, 291, 297; Herrera Oria *et al.*, p. 337.
16. Fernández Duro, ii. 298, 301, 473; Herrera Oria *et al.*, p. 289; A. Crinò, 'La partecipazione di un galeone del Granduca di Toscana', *Archivio storico italiano*, 142 (1984), 605.
17. Herrera Oria *et al.*, p. 292; Hume, iv. 466–8.
18. Laughton, i. 305, 364, ii. 127; Fernández Duro, ii. 311; Maura, pp. 265–6.
19. Herrera Oria *et al.*, pp. 287–8; N. Alonso Cortés, *Miscellanea valisoletana*, 3 (Valladolid, 1921), 32–4; J. M. Lope Toledo, 'Logrono en el desastre de la Armada invencible', *Berceo*, 17 (1962), 234–7.
20. Fernández Duro, ii. 459–64.
21. Laughton, i. 258, ii. 96–7.
22. Laughton, ii. 138–9, 141.
23. Herrera Oria *et al.*, pp. 291–2, 300–1, 344; Laughton, ii. 183–4, 212; Lyell, p. 166; Fernández Duro, ii. 470–3.
24. Hume, iv. 466–8. The assessment of the Armada losses which follows is based on Fernández Duro, ii. 330, modified by the facts given in the next chapter.

Chapter Nine

1. Maura, p. 257; Fernández Duro, ii. 126; Laughton, i. 201, 218, 258, 312, 314; Hume, iv. 355–7.
2. Laughton, ii. 299; Naish, p. 84; Fernández Duro, ii. 314.
3. Fernández Duro, ii. 320–1; *Certain Advertisements out of Ireland concerning the Losses and Distresses happened to the Spanish Navy upon the West Coasts of Ireland* (London, 1588); Martin, pp. 23–135. The rest of this chapter is based on *Calendar of State Papers: Ireland*, ed. H. C. Hamilton (London, 1885), iv. 26–77.
4. Martin, p. 122.
5. N. Fallon, *The Armada in Ireland* (London, 1978), 151–3; Fernández Duro, ii. 272, 287; Coco Calderón, p. 36; cf. *Calendar of State Papers, Carew MSS*, eds. J. S. Brewer and W. Bullen (London, 1868), ii. 451. G. Mattingly seems to have had access to a source which has eluded me: *The Defeat of the Spanish Armada* (Harmondsworth, 1965), 389.
6. Fernández Duro, ii. 398.
7. Martin, pp. 189–244; Laughton, ii. 273–4.
8. Martin, pp. 137–87; Fernández Duro, ii. 289–92.
9. Sténuit, pp. 112–27; Hume, iv. 500.
10. Fallon, *The Armada in Ireland*, pp. 48, 195–6.
11. Fernandez Duro, ii. 337–70. Cuellar's story has often been told, most notably by J. A. Froude, *The Spanish Story of the Armada and Other Essays* (London, 1892).
12. Hume, iv. 463; *Calendar of State Papers, Carew MSS*, ii. 472.
13. J. de Courcy Ireland, 'Ragusa and the Spanish Armada', *The Mariner's Mirror*, 64 (1978), 257–60; Coco Calderón, p. 39; Hume, iv. 456–9; Fernández Duro, ii. 419.
14. Laughton, ii. 277–8, 290, 293.
15. Fernández Duro, ii. 476.

Glossary

Ship types

argosy	the largest type of merchant ship, its name perhaps a corruption of *Ragusino* (Ragusan, that is, of Dubrovnik).
carrack	any type of large merchantman.
flyboat	small, fast, shallow-draughted ship for coastal warfare.
galleass	large warship powered by both oars and sail.
galleon	large, relatively long, relatively low-built, sail-powered, ocean-going warship.
galley	low, relatively shallow-draughted, oar-powered ship.
hulk	large, short, round-hulled or tub-shaped merchant ship.
Levanter	any large sail-powered ship of Mediterranean build.
pinnace	small, fast, light, usually two-masted vessel; used as messenger-ships.
tender	any small ship attending on a large one, particularly distinguished from a pinnace by the latter's essential speed.

Other nautical terms

bear room	to stand off; to alter course, or slacken or increase sail, so as to create more sea-space relative to other shipping.
bowline	rope passed from the edge of the sails on the windward side to the opposite side of the bow. By 'hauling on the bowline' sailors can keep the sail steady when sailing close to the wind. The ship is then said to be 'close-hauled'.
bowsprit	large boom projecting from the stem of a ship.
carling	step of a mast.
close-hauled	see bowline.
give luff	present a broadside to the enemy from a windward position in the course of an about-turn.
gromet	ship's boy.
haul on the bowline	see bowline.
luff	windward side of a ship; 'to luff' is to bring one's prow round to the wind.

run before the wind   to sail where the wind takes one.

tack   to sail obliquely against the wind, first to one side, then the other;
                  tacks are ancillary lines run from the windward edge of a sail (cf.
                  bowline).

warp   to tow on a rope or 'warp' in a calm or against the tide.

weather-gauge   a position to windward of other shipping.

Acknowledgements

The author and publishers wish to thank the following who have kindly given permission to reproduce illustrations on the pages indicated;

Archivio di Stato, Florence, p. 149
© Bibliothèque Royale Albert 1er, Brussels (Cabinet des Estamps), p. 131
Bodleian Library, pp. 47, 73
The British Library, pp. 2, 7, 13, 68, 91, 92, 108, 110, 111, 112, 113, 123, 142, 143, 147, 172, 175, 186 top, 199, 207, 225, 228, 229, 231, 266
The Governing Body of Christ Church, Oxford, p. 21
B. J. Harris, Oxford, p. 272
The Controller of Her Majesty's Stationery Office for Crown copyright material in the Public Record Office, pp. 25 (MPF 208), 96 (MPF 318)
Milo Kovač, Dubrovnik, pp. 244, 245
Kunsthistorisches Museum, Vienna, p. 100
The Master and Fellows of Magdalene College, Cambridge, pp. 6, 18, 148, 156
Musées Royaux des Beaux-Arts, Brussels and © A.C.L. Brussels, p. 84
Museo del Prado, Madrid, pp. 38, 39
Museum de Lakenhal, Leiden, p. 35
The Trustees of the National Gallery, London, p. 37
National Maritime Museum, London, pp. 32, 61, 159, 176, 179, 186 bottom, 204, 212, 264, 275
National Portrait Gallery, London, pp. 3, 94
The Collection at Parham Park, West Sussex, p. 221
Patrimonio Nacional, Madrid, and Mas, pp. 138, 139
Rijksmuseum, Amsterdam, p. 117
Royal Commission on the Historical Monuments of England, p. 236
St Faith's Church, Gaywood, King's Lynn, p. 118
St Peter's Church, Tiverton, p. 235
The Marquess of Salisbury, p. 81
The Marquesa de Santa Cruz and The National Maritime Museum, London, p. 9
Shetland Museum, p. 146
The Marquess of Tavistock and The Trustees of the Bedford Estates, p. 240
Tiroler Landesmuseum Ferdinandeum, Innsbruck, p. 187
The Ulster Museum, pp. 31, 51, 53, 58, 63, 70, 145, 169, 254, 259
Board of Trustees of the Victoria and Albert Museum, p. 219
Weidenfeld and Nicolson Archives, p. 105
The Worshipful Society of Apothecaries, p. 192

Picture research by Sandra Assersohn

Index

Notes. All references are to Spain and England, unless otherwise indicated. Sub-entries are in alphabetical order except where chronological order is more relevant. Page numbers in *italics* refer to illustrations and/or their captions. There are occasionally textual references on the same pages.